HOW CONTEXT MATTERS

HOW
CONTEXT MATTERS

LINKING ENVIRONMENTAL
POLICY TO PEOPLE AND PLACE

George Honadle

KUMARIAN
PRESS

This volume is dedicated to the memory of

Tom and Roberta Worrick

and all others who have lost their lives
or made other great sacrifices
in the quest for a
sustainable
future

How Context Matters: Linking Environmental Policy to People and Place

Published 1999 in the United States of America by Kumarian Press, Inc.,
14 Oakwood Avenue, West Hartford, Connecticut 06119-2127 USA.

*Production and design by The Sarov Press, Stratford, Connecticut.
Index by Linda Webster.
The text of this book is set in Sabon 10/14.*

Printed in Canada by AGMV Marquis.
Text printed with vegetable oil-based ink.

∞ The paper used in this publication meets the minimum requirements
of the American National Standard for Information Sciences—Permanence of
Paper for Printed Library Materials, ANSI Z39.48–1984.

Library of Congress Cataloging-in-Publication Data
Honadle, George.
    How context matters : linking environmental policy to people and
place / George Honadle.
       p.   cm.
    Includes bibliographical references and index.
    ISBN 1–56549–105–X (cloth : alk. paper). — ISBN 1–56549–104–1
(pbk. : alk. paper)
    1. Environmental policy. 2. Environmental protection. I. Title
GE170.H66 1999
363.7'05—dc21   99–44133

08 07 06 05 04 03 02 01 00 99   10 9 8 7 6 5 4 3 2 1

First Printing 1999

Contents

LIST OF FIGURES

PROLOGUE

D o you take your coffee black or with cream and sugar?" the server asks as he hands the customer a menu. This scene occurs in uncounted restaurants many times each day. And yet, each time I encounter it, I wonder how many people find it curious.

It is based on an assumption that coffee is consumed independently of the food that accompanies it—each individual is expected to prefer coffee prepared in a certain way. That is the way it is supposed to be. Anything else must be odd. But, to me, an unwavering mixture is the oddity.

I cannot say I will have coffee, or select its form, until I see the whole picture. For example, if I have a sore throat, I may want cream and sugar to overcome the harshness and to provide a sweet coating; if I am eating something very sweet, such as pecan pie, I will drink it black to cut through the sweetness; if my breakfast includes dry bacon on the side, I will put cream in the coffee to soften the bacon aftertaste. Sometimes I will charge the caffeine with sugar alone to hasten the rush of alertness, and other times I will drink it sweetened with honey. In the warm months I may drink iced coffee, and on rare occasions, after a meal, I may take black coffee and pour thick cream over the back of a spoon to create a layer of cream on top to sip the coffee through. Most often I don't drink coffee at all. But never do I choose my coffee regime before deciding what I will eat.

Most people, however, do not seem to see it this way—they drink their coffee the same way no matter what they eat along with it. This also symbolizes the way many people approach life in general—the search is for the preferred way to do something. Once it is found, that is the approach that is used ever after.

This also applies to professional agendas. Advocates of particular positions become identified with those positions, solutions for problems become standardized, and organizational success is equated with introducing a particular solution. As with coffee, the world is divided into black and white.

But this way of thinking may be part of the problem. In the area of environmental policy it may blind us to the fact that what works under one set of circumstances may not work under others. There may be no simple solutions for many environmental problems, and the way we search for solutions may itself constitute a problem—the coffee alone should not be the focus.

I encountered just such a situation a few years ago. The task was to purchase a refrigerator. I wanted the operating costs and the energy use to be as low as possible. For assistance in this effort I turned to the literature on green products.

Guidance on many activities is offered in a widely available environmental publication on green consumerism. That pamphlet indicated that it was better to buy an over-under configured refrigerator-freezer than a side-by-side one. For any given cubic footage the over-under was supposed to be more efficient.

I discovered that it wasn't necessarily so. The advice was good only if you considered the total volume of the combined units (the way they were advertised). But if you were interested in refrigerator space, the generalization did not hold. Over-unders tended to have a greater percentage of the total volume devoted to freezer space. Since we lived on a farm and had additional freezer space, we needed less freezer and more refrigerator in our kitchen unit.

If we kept the refrigerator space constant, we were comparing a smaller size side-by-side to a larger over-under, and both the cost and energy use of the side-by-side were better. Thus, the green consumer advice was sometimes erroneous. The generalization did not hold when the objectives and situation varied. Indeed, environmentalists, too, often saw the coffee and missed the context.

This is especially disturbing because, by definition, an ecological perspective should be a contextual perspective. It should encompass the entirety of a setting, not just an object in it. But, too often, it does not. Even well-trained professionals fail to see differences among objectives and settings and they reach conclusions and make recommendations that lead to poor decisions.

To rectify this, we need to focus directly on context. This is especially true in the arena of environmental policy formulation and implementation because this is a subject where the context often is the source of the problem and therefore it should be the target of the policy. But to revisit context effectively, we need to refine our perception of it.

That is the purpose of this book—to clarify those elements of context that affect how policies work and to show the linkages among public policies, human behavior, and the vigor of natural settings. It presents the context hypothesis.

ACKNOWLEDGMENTS

A book is simply a product that marks a point in a process—a process of experiencing, learning, and grappling with ideas and information. This volume marks such a point for me. It gathers the artifacts of many conversations, work assignments, collegial meetings, and hours pouring over written documents and concentrates them on a topic—crafting policies for the achievement of sustainable societies. To say that I owe much to many people who contributed to that process is an understatement. To give them credit for helping me and facilitating this product is the least that I can do. Just as there is a common saying in Africa that it takes a village to raise a child, so too those who have tried know that it takes many people to make a book. And just as it is impossible to remember the role of every villager in the nurture of the child, it is also impossible to credit everyone who influenced my thinking and affected the nature of this volume. So, I apologize to anyone that I have missed in my attempt to acknowledge the help I received.

One form of help is financial. It is easier to write when the bills are paid and the travel and time are funded. My support on this front came from two sources. The first is my wife Beth, formerly a Professor in the Department of Applied Economics at the University of Minnesota and now Professor of Political Science and Director of the Center for Policy Analysis and Public Service at Bowling Green State University. The second was from funding to the University of Minnesota through a collaborative agreement between the U.S. Agency for International Development (USAID) and the Midwest Universities Consortium for International Activities (MUCIA). An early version of this publication was supported in part by the Environmental and Natural Resources Policy and Training (EPAT) Project funded by USAID. I thus owe a great debt to USAID, the University of Minnesota, and MUCIA for support and encouragement during the research and writing that led to the report

Context and Consequence, which became a springboard to this work. A second study was commissioned, through EPAT, by the Africa Bureau of USAID. The product of that study, titled *The Problem of Linear Project Thinking in a Non-linear World* was an examination of programming implications of the interaction among agricultural, population and environmental programs in Africa. It also shaped some of the thinking in this book. For their assistance in these efforts, I would like to identify Hans Gregersen, Richard Skok, Al Sullivan, Alan Ek, William Fenster, Delane Welsch, Julie Borris, Linda Lamke, Janelle Schnadt, Will Candler, Nick Poulton, Ken Baum, Norma Adams, Tony Pryor, Michael Rock, Scott Grosse, and Paul Phumpiu for special mention.

A second form of help is experiential. The numerous organizations that engaged my services over the years helped me to build my experience base to nearly one hundred consulting assignments in twenty-seven countries. I thank organizations such as the World Bank, World Wildlife Fund, Agency for International Development, United Nations Development Programme, African Development Bank, Asian Development Bank, Minnesota Department of Natural Resources, Minnesota Environmental Quality Board, Association of Minnesota Counties, and the McKnight and Northwest Area Foundations for providing me with such opportunities. Consulting firms such as Development Alternatives, Inc., Abt Associates, Management Systems International, Lauren Cooper Associates, and Associates in Rural Development also provided contractual mechanisms and team environments that made some of the field work possible.

The chance to serve on the Minnesota Roundtable on Sustainable Development and on the Citizens Advisory Task Force for the Chisago Electric Transmission Line Project also exposed me to the local politics of environmental progress. And where there is politics, there is stress, and where there is stress, there are often innovators and sometimes even heros. Various members of the Roundtable qualified as innovators, and some of the members of the Task Force certainly qualified as heros. I thoroughly enjoyed getting to know and working with all of them. They will recognize many of the perspectives and some of the examples presented in this book.

The actual massaging of a manuscript and production of a publication also requires the efforts of many people. At Kumarian Press I am most indebted to the early support and confidence of Krishna Sondhi. And later, during the completion of the project, the guidance and assistance of Linda Beyus and a reviewer were most appreciated. Their efforts added greatly to the final product.

Frances Korten and Paul Harrison also deserve special mention. They reviewed my reinterpretation of their work, corrected some of my mistakes, offered detailed comments on my draft, and supported both my conclusions regarding their experiences and important contributions as well as the potential of the context perspective to increase our understanding of them. Their efforts and encouragement are greatly appreciated.

Although the views, interpretations, opinions, and errors are mine alone and should not be attributed to any other source, many colleagues offered encouragement, criticism, ideas, and discussion that led to issues covered in the book—Sandra Archibald, Derick Brinkerhoff, Thomas Carroll, Michael Cernea, Clement Dorm-Adzobu, Joel Heinen, Edward Karch, Dennis King, Rudi Klauss, Scott McCormick, John D. Montgomery, Emilio Moran, K. Robert Nilsson, James Nations, Shanaz Padamsee, J. Kathy Parker, Rogerio Pinto, Dan Ray, Bruce Rich, Peter Sand, Rodney Sando, Margaret Shannon, Stephen Schwartzman, Andrea Silverman, Jerry Silverman, Virginia Stark, Karl Stauber, Barbara Toren, Paul Toren, and John Wells all provided me with helpful perspectives at different stages in the book's evolution. My students in a course on "Natural Resources Policy and Management in Developing Countries" also added questions and comments. And many professionals, civil servants, NGO staff, and villagers in Africa and throughout the globe made me see things more clearly. But, again, these people bear no responsibility for any errors of omission, commission, interpretation, characterization, presentation, or argument. Such responsibility is mine alone.

1

INTRODUCTION:
BEYOND HOMOGENOUS THINKING

The view from my Minnesota office included the salt lick next to the creek and three deer approaching it. The snow on the ground, the squirrels in the trees, and the bright blue of the sky with the sun sinking behind the woods, all symbolized a world separate from the computer and fax machine by my side. And yet it was not. That world outside was the focus of many battles, and these machines were weapons in the arsenal that was being mobilized to defend it from a human onslaught.

But there are many versions of that world. Over thirty years ago, as a Peace Corps Volunteer in Malawi, I stood watching a similarly beautiful sunset. I was asked what I was doing. Then, when I answered, my friend shook his head and smiled, not believing that I would find anything remarkable in such an ordinary event. Indeed, in his language, the word for "beautiful" could not be applied to a sunset.[1]

Like my village friend, when most people discuss problems of interest to them, they adopt a language and a set of assumptions that they share. This protects them from the need to recognize radically different world views. The search for universal solutions (such as a free trade regime or a market-based economy or an endangered habitat act or a ban on ivory sales) supersedes discussion of uniqueness. The answers that are sought are generic, not contingent. That is, there is little entertainment of the idea that a set of policies or actions can produce radically different consequences under different situations. But they can, and they have.

This is especially true for environmental policies. Local economic circumstances, historical trends and socio-political dynamics all can combine to alter outcomes and to narrow the options available for policy implementation. Problems arise when policy analysts approach implementation by simply replicating the trappings of strategies used in other contexts.

This study begins from the premise that we need to understand more fully what makes situations different and that we need to learn more specifically how those differences cause similar policies and strategies to produce different results. But the published literature deals more in broad prescriptions than in situational typologies. Thus, we need to construct new lenses for viewing the interactions between contexts and consequences. Before that process can begin, however, we should know where we are, how we got here, and why it is important to move in new directions.

RECOGNIZING THE IMPORTANCE OF CONTEXT

Why is context important for environmental policy? Contextual thinking is important for three tactical reasons:

- *first*, a policy that is appropriate in one locale may lead to disastrous results in another—context is important for determining substance;

- *second*, context influences the processes that can be used to formulate policy—without contextual sensitivity, effective policies may never be developed; and

- *third*, contextual maps are needed to execute policy—context stands in the way of the transformation of pronouncement into performance, and context changes over time (Ankney, 1996).

Given the importance to the human species of enacting environmental policies all over the globe, the need to understand context cannot be overstated. There also is another, overarching, strategic reason why context is so important:

- *fourth*, we are running out of time.

Either the call for a new environmental ethic (Orr, 1992) or faith in the instincts of the next generation (Chertow and Esty, 1997) pervades the concluding sections of many of today's publications dealing with the promotion of world greening. But instilling loftier values, or counting upon more enlightened views, in future generations may come too late. We need approaches to slow, halt, reverse, and replace human behavior in the short run. The immediate need is to identify how to reach actors who respond to different incentives and who locate themselves in different realities than those advocates of green societies (Lewis, 1992). This requires an understanding of situational diversity on the part of those promoting new policies.

Our focus is determined by the tools we have, rather than by what we need, when what we need is missing from the toolkit. It is much like the man who lost his wallet and searched for it under the lamppost because it was lighted there, even though that was not where he thought he lost it. We all give lip service to the need to understand context, but we stop short of looking directly at it. We go where the light is instead of constructing new lampposts.

The lights we shine on natural resource policy may brighten the view of nature, but they seldom illuminate why or how people do what they do to nature. Policies are identified more with the natural resources that they are designed to protect than with the different intermediate consequences and human behavior changes that they are expected to produce.

This is a key point. Natural resource policies are usually grouped by the physical conditions they address, such as watershed rehabilitation, erosion control, or pollution abatement, or else they are described as regulatory versus market-based. In neither case are they categorized by the different types of human behaviors that they attempt to evoke. But, from an implementation perspective, it is the different behavioral expectations that are important.

As long as we treat policies as an undifferentiated lump we will not be able to see that when different sets of intended consequences combine with different contexts, they produce different outcomes. And we need different strategies to achieve similar outcomes in different settings. Intuitively we know this, but international donors and conservation organizations seem to remain wedded to their preferred strategies and policy solutions. The programs of particular international organizations are easily recognizable as one moves from country to country. Both development policy and environmental strategy have become bureaucratized, and bureaucratization requires standardization. As a result, we have homogenized the world to make it easier for management. But, if we are to break loose from our blinders, we had better understand how we got where we are.

COMING TO HOMOGENEOUS THINKING

In the 1960s and 1970s the professional field of international development was in ferment. Contending theories of balanced versus imbalanced growth, import substitution versus export promotion, stages of economic growth versus dependency theory, enclave support versus broad-based participation, rural investment and agricultural production versus industrialization, planned economies versus market-based strategies, comparative advantage and commodity dependence versus diversification, and infant industries versus free trade were among the macro alternatives debated and promoted. At a micro level, theories of optimizing peasants, backward bending supply curves, cultural constraints and consciousness raising, bypass strategies using parastatal enterprises to force savings and efficiency, imported expertise versus indigenous knowledge, appropriate technology versus technological leap-frogging, and blueprint versus process models of project design were bandied about. But by the mid-1980s this all changed.

The change occurred as a result of converging incidents and experiences. First, the performance record of many third world economies, especially in Africa, left much to be desired. Tolerance for overly bureaucratized development strategies wore thin. Second, the World Bank published a report on Sub-Saharan Africa that suggested open economies with market-based strategies were more successful than more planned and controlled economies (World Bank, 1981). Third, international debt burdens reached staggering proportions. And fourth, the United States adopted a policy platform that promoted deregulation, free trade and a preference for non-governmental, market-oriented development strategies. Debate was stifled as solutions were imposed.

This resulted in resources, through multilateral and bilateral agencies, being committed to a particular strategy. Parastatals were out and non-governmental organizations (NGOs) were in.[2] Planning was out and free markets were in. Delivery systems were out and getting the prices right was in. Projects were out and policy dialogue was in.

Some of this approach was supported by field evidence. The development management literature had long observed that national policy constraints could make local project-level improvements futile—grass roots successes almost never transformed into large scale results. Centralized administrative structures proved generally incapable of spawning the creative environments necessary for development. And the evidence of farmer response to farm-gate prices dispelled any remnants of backward bending supply curves.[3] The new orthodoxy of development became a standard sequence—first create the right policy environment, and then follow through with project investments. Because there was a claimed vision of how to do it, tolerance of national differences declined. There was one perceived historical path to success. One outcome of such thinking was policy dialogue bent on imposing private sector led development. This amounted to avoidance of contextual considerations.

While the development community was shifting into a much more homogeneous and commercial view of the world, the conservation community became both more financially secure because of reactions to that world view and more strident in response to it. And since this was largely a community of non-governmental organizations (NGOs), it gained a stronger foothold on development. In contrast to the public sector, conservationists found themselves part of a growth market—the non-governmental sector.

But conservation strategies were also changing. The evidence of natural resource depletion was mounting daily, and the threat this posed to development became more and more irrefutable. And due to the march of human numbers, the need for development to ease the pressures on fragile ecosystems became more apparent. Indeed, protection and development became more obviously necessary complements as poverty, population growth and environmental degradation reinforced each other (Leonard and others, 1989; Timberlake, 1985). Sustainability then became the battle cry as connections among ecosystem health and social system robustness became more clear.

Awareness of these connections propelled new conservation views to the fore. What had been adamant nature protectionism became tainted with such themes as buffer zones and ecotourism. Natural scientists began to accept that the social sciences, and even economics, might have a role in the protection of natural areas. Whereas the private sector had been seen as extractive and destructive, it now surfaced as a force to be harnessed for protection and sustainable use. A new alliance was created between main line conservationists and conservative development critics. Both could agree on the need to rein in multilateral organizations such as the World Bank and regional development banks. And both were used to thinking in terms of "one best way."

But the more experience we gained in promoting sustainable technologies, restructuring economies, and revitalizing natural resources, the more it became obvious that the answers were not so readily available or so universally applicable. Birth control incentives that worked in India failed in Kenya. Policies that promoted competition in Asia supported monopoly and shadow markets in Africa. The problem of food production in water rich areas was far different from that in places where irrigation is an impossibility. Decentralized approaches in large and populous states like Nigeria or Brazil bore no resemblance to the meaning of decentralization in small, scattered island nations like Vanuatu or the Comoros. An approach to the ivory trade that was adopted in East Africa was reviled in Southern Africa—and each may have made sense under the different circumstances that prevailed in those two places. (Bonner, 1993; Thomson, 1986; Rose, 1992; Dobson and Poole, 1992)

What does remain constant among these illustrations, however, is the tendency of people to assess their options in terms of what they face immediately. The enduring lesson of a large body of literature called "rational choice" is that people do respond to the web of incentives and disincentives that surrounds them, and there is great variability in those webs from place to place throughout the world (Bates, 1981; Ostrom, 1990; Popkin, 1979; Russell and Nicholson, 1981). Moreover, this applies both to the social settings encountered by villagers and to the organizational settings encountered by bureaucrats and policy-makers (Honadle and VanSant, 1985; Ostrom, Schroeder and Wynne, 1993). The key is finding strategies that fit specific circumstances.

This was thrust into the foreground during a field assignment in Kenya in the mid-eighties. I was involved in designing a project to support intermediary organizations in the agricultural sector. This was the heyday of the faith in the power of prices to inspire farmers to produce more of any crop. Indeed, prices were important. But interviews with producers revealed that in two specific commodities price was not the primary concern. Milk producers were willing to accept even lower prices if the risk of spoilage was shared with the purchaser. And cotton farmers were willing to get lower prices if honest weighing and accurate grading could be guaranteed. In fact, they had no faith that they would be better off with higher prices. The world was not as homogeneous as was commonly thought. Today we are sitting on the cusp of a past that promoted homogeneity and a future that will force us to accept diversity. The question is how to move our thinking from the one to the other.

MOVING TOWARD SITUATIONAL THINKING

In the closing decade of the twentieth century, the changing demographics of the North American continent have propelled the term "cultural diversity" to a position of prominence in education, news reporting, politics and business. Simultaneously, world events from Africa to eastern Europe to the orient give testimony to the desire of diverse ethnic groups to achieve political autonomy and historical legitimacy. Political empires are fragmenting and people are separating into smaller scale units at the same time that industrial empires are expanding and large scale interactions are becoming increasingly identified as the source of environmental problems. Industrial emissions in North America cause acid rain in Northern Europe; airborne chemicals from China travel to Texas; demand in the metropolis threatens biological diversity in the Amazon, and that diversity is necessary for the survival of the metropolis. Planetary connections are becoming known at the same time as the virtues of small scale and variety are being touted in both social and scientific arenas.

Natural scientists extol the virtues of species, genetic and ecosystem diversity (Wilson, 1992; Baskin, 1998, Tilman, Wedin and Knops, 1996) and the importance of context for understanding scientific connections and evolutionary processes (Cohen and Stewart, 1994; Drury, 1998; Zimmer, 1998). Business school faculty go so far as to assert that competitive advantages in commerce derive from differences (Porter, 1990), economists argue that our tools for analyzing the world too often ignore the differences that define third world socio-economic dynamics (Bromley and Cernea, 1989; Anheier, 1992), and anthropologists insist that development processes that drown traditional cultures are the cause of development failures and the elusiveness of sustainability (Anderson, 1990; Anderson and Huber, 1988; Clay, 1988; Hecht and Cockburn, 1990; Nabhan, 1997; Richardson, 1976).

Simultaneously, a new school of environmental historians is emphasizing how the past human effect on the natural setting, the physical landscape, and the present senses of "place" should be advanced to center stage in any discussion of natural resource policy (Power, 1996; Wilkinson, 1992; Cronon, Miles and Gitlin, 1992). Clearly it is time for context to become the focal point in discussions of sustainable development and environmental policy reform.

We do recognize that problems have different roots in different places. For example, the causes of tropical deforestation can be traced back to two major factors—human population increase and the penetration of world market forces into third world economies. But the ways that these two causes manifest themselves in different places vary considerably, as shown in Figure 1–A.

Policies and programs cannot be based on generalizations, they must fit specific local conditions. The same is true of implementation strategies. Reform efforts must be custom-tailored to the time and the place. But for that to happen, we need to know what the elements are that influence strategy choice and effectiveness.

True diversity implies variety of context. A species is different because it has adapted to different surroundings. A culture is unique because it has developed its own response to different historical and physical conditions. Contextual thinking requires appreciation of variety and utilization of differences. Indeed, it pushes the focus on differences among settings to the foreground.

This study will focus on how different settings and objectives limit the ways that policies produce results, how they impose different constraints on

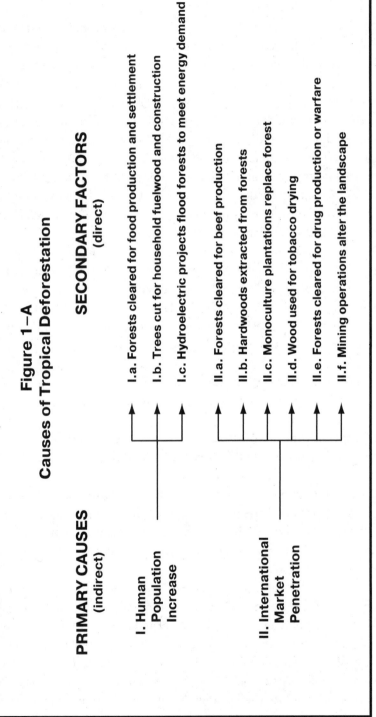

**Figure 1–A
Causes of Tropical Deforestation**

PRIMARY CAUSES
(indirect)

SECONDARY FACTORS
(direct)

I. Human Population Increase

I.a. Forests cleared for food production and settlement
I.b. Trees cut for household fuelwood and construction
I.c. Hydroelectric projects flood forests to meet energy demand

II. International Market Penetration

II.a. Forests cleared for beef production
II.b. Hardwoods extracted from forests
II.c. Monoculture plantations replace forest
II.d. Wood used for tobacco drying
II.e. Forests cleared for drug production or warfare
II.f. Mining operations alter the landscape

Source: George Honadle. 1989. *Putting the Brakes on Tropical Deforestation: Some Institutional Considerations,* Washington, DC: Agency for International Development, p. 4.

the strategy choices available, and how they often present unanticipated opportunities. We will move people and place to center stage in the drama associated with the implementation of natural resource, environmental, conservation, and sustainable development policies.[4] But both literature and experience suggest a major weakness in the prevailing perspectives on policy formulation and implementation—no one is able to identify the key elements of context that affect implementation strategy and impact. Although there may be an underlying order, it has not yet been charted. Even those who agree that for policies to work "it all depends" cannot point to what it depends upon. That is because there is no map of context. We have to draw our own.

Like most maps, the one we draw will be an approximation, rather than an exact reproduction, of the terrain we expect to traverse. It will not be an algorithm that always leads us to the perfect policy implementation strategy. But it should still have heuristic value—it should help us to anticipate difficulties, to plan ways around them, to understand the idiosyncracies of the landscapes we enter, and to reach our destination in better condition than we would otherwise. It should help us to avoid major mistakes when we select environmental policy alternatives and it might help us to achieve a sustainable future.

NOTES

1. In Chitumbuka (the language of the Tumbuka people of Malawi and Zambia) the word for "beautiful" is *kutowa*. This is the infinitive form of a verb—beautiful is a verb describing a process of being and not an adjective attributing a characteristic to an object. A person, therefore, is depicted as "being beautiful" (*akutowa*) when the word is used. Non-human objects or creatures cannot engage in the process of being beautiful and thus the grammar of Chitumbuka did not allow the sunset to be depicted as "beautiful."

2. NGO stands for Non-Governmental Organization, but it actually refers only to non-profit private organizations or community-owned organizations.

3. In the 1940s and '50s experience in colonial areas suggested that many people would produce a cash crop up to a level where they reached a certain income and then stop. From this point on, time to do other things was more important to them than extra income. This phenomenon was described by economists as a "backward bending supply curve" because when supply was plotted against price, the supply curve turned backward at the point where the price was high enough for the producer to achieve the targeted income level.

4. Distinctions can be made among these types of policies: A *natural resource policy* is any policy directly dealing with the management, distribution, stockpiling,

extraction, use, depletion, or protection of any stock or flow of natural commodities; An *environmental policy* is any policy stressing the conservation, preservation, protection or sustainable use of any natural commodity, landscape, habitat, organism, or community or any policy affecting human behavior that directly or indirectly alters the stock, flow or condition of any natural commodity, landscape, habitat, organism, or community; A *conservation policy* is any policy intended to conserve, preserve, or protect any natural or semi-natural landscape, habitat, community, organism, or genetic material; A *sustainable development policy* is any policy intended to make economic and social activity supportive of environmental resilience and able to endure through many generations into the indefinite future. But we will use "natural resource policy" and "environmental policy" in a generic sense to encompass the range of policy initiatives concerned with the health of natural areas.

2

DIFFERENT DEMANDS: INTENDED CONSEQUENCES OF NATURAL RESOURCE POLICIES

Natural resource policies are not aimed at natural resources—they are aimed at people. Nature does not react to human policy, but it must contend with human behavior. Since policies are intended to guide human behavior they must be firmly grounded in the world experienced by the actors whose behavior is the object of the policy. If they are not well-grounded, the chances for success plummet.

The call for policies to be based on "good science" echoes around the meeting rooms, hallways, and offices of NGOs, government agencies, and academic institutions. But the call for social soundness is muted at best. The assumption is that the arguments of science will win out if only they are heard. But experience suggests otherwise.

The Ecuadorean homesteader who fells trees to establish a legal claim to the land may see the wisdom of keeping the trees, but as long as the legal claim requires that the trees go down he will continue to cut. Likewise, the Filipino fisherman who sees his harvesting practices destroying the coral reef knows that his practices are not sustainable and they are dooming his own future, but he needs the money now. Clearly, policies must support alternative behavior in the short run and not just appeal to a sense of ecological justice or to an awareness of the long-term consequences of present activity.

Nevertheless, when attempts are made to extract lessons from the social science or management literature, they often fall short because of homogeneous thinking. Researchers commonly assume, or conclude (Brinkerhoff, Gage and Yeager, 1992, for example), that *clear* objectives are all we need without examining whether *different* objectives are important.

But an architect cannot design a building without knowing its intended use—different uses dictate very different designs. Nor should we expect policies to be independent of their objectives. Even so, all previous studies seem willing to say is either that clear statements of use lead to good designs (that is, a good implementation process is characterized by clear policy objectives), or that programs that begin with clear, simple objectives tend to become encumbered with more complex and fuzzy sets of multiple objectives because the early ones fail to capture the essence of the problem (Mather, 1993).

Clearly, this is not enough. We need to know how different uses (intended consequences) dictate a need for different designs. We need to distinguish among the different behavioral objectives of environmental policies if we are to improve our understanding of the strengths and weaknesses of alternative implementation strategies.

Some studies have been on the verge of categorizing policies based on behavioral objectives, but have failed to do so. For example, one literature review of natural resource policy implementation in Africa states:

Kenyan Wildlife policy aims to preserve biodiversity by changing attitudes and behaviors of communities adjacent to parks and reserves, with a focus on economic incentives for conservation. In Lesotho, rangeland policy reform involves reallocations of resources and authority between public and non-governmental sector organizations involved in rangeland management. Reforms in forestry policy in Senegal, Mali and the Gambia target community-level behaviors in combination with changes in the mission and operating procedures of government forestry departments. Among the policy initiatives in Uganda and Madagascar is the development of an overarching planning framework for environmental action that specifies targets, indicators, timeframes, and roles and responsibilities. (Brinkerhoff, Gage and Yeager, 1992: 3)

But then the ball is dropped and the study goes on to look at policy implementation in terms of characteristics of the donor-funded effort—characteristics such as the presence or absence of training, studies, equipment and so forth.[1]

A more penetrating perspective is needed to further the state of the art. We must identify different categories of consequences to analyze strategic needs, and we must specify the implementation implications of different intended consequences.

Two groups of intended policy consequences are readily discernable. The first is targeted consequences—where impact is expected on certain people, resources, places, or organizations. The second is systemic consequences—

where impact is much more diffuse, and perhaps more pervasive, in a society. Both categories of intended consequences are noted below.

TARGETED CONSEQUENCES

A wide range of behaviors may be the subject of change attempts. Some change strategies will involve motivating people to do things that are new. Others just will involve getting them to stop something they have been doing.

Reducing Damaging Behavior

Policies, legislation, and treaties dealing with natural resources often try to stop people from doing things. Banning international trade in endangered species, stopping the application of hazardous pesticides, halting the harvesting of whales, blocking the cutting of trees, limiting the discharge of airborne pollutants, prohibiting the dumping of waste in streams, or keeping builders or settlers out of fragile ecosystems are just a few examples.

Stopping damaging behavior implies monitoring human behavior, and the condition of the resource, plus police action. Specific activities are forbidden and a cadre of enforcers ensures compliance with the ban. The focus is on penalties for non-compliance. Negative reinforcement and public display of punishment characterize this approach. A punitive mentality prevails.

Behavior blocking is a costly and conflict-ridden approach that assumes the worst about human nature or human circumstances—only countervailing force will prevent damaging behavior. People are the problem.

Negative reinforcement is well-grounded in psychological experimentation (Skinner, 1972) and social control. Because it is punishment-oriented, it benefits from social distance between controller and controlled, such as with police and felons or with game wardens and poachers. In fact, difficulties in the performance of third-world agricultural cooperatives often result from simultaneously needing to prevent deleterious behavior and needing to confront a social equal and co-member. And extension agents charged with both supporting farmers through the provision of technical information and controlling them by collecting loan balances have similar problems. It is very hard to do both in the same action or have them done by the same people.

Clear objectives also are needed to block behavior. The clarity of the objective is one of the key elements in promoting success. People must know precisely what they may not do. The difficulty and cost of enforcement increase dramatically when fuzzy rules require interpretation of each action and allow great discretionary latitude.

And negative reinforcement requires a clear and direct link between behavior and punishment or else it does not work. The swiftness of retribution is important to deterrent power because it clarifies that link.

There is also the question of exactly what behavior is discouraged by a policy. In at least one American state, for example, it is illegal to sell detergents containing phosphates, but it is legal to use them. A decision was made to target merchants rather than households, presumably because it was politically and administratively easier to monitor and prosecute them. In other cultural settings, or situations where most of the population resided near a border, however, such an approach would simply create a shadow market.

Behavior blocking is probably the most commonly perceived type of consequence associated with environmental policy. But an important, and increasingly common, approach in third world communities is the mobilization of effort to restore a degraded resource and regain environmental services such as shade, soil retention, and cattle fodder (Harrison, 1987).

Restoring a Resource

Getting people to restore a degraded resource is not the same as stopping the degradation in the first place. The type of social organization needed is very different.

Restoration may be an individual, organizational, or family activity, but in third world settings it is commonly a community effort. It is not a matter of police versus trespassers, but rather it is community leaders mobilizing temporary efforts—work crews or task forces. And getting someone to commit time, labor, and other resources is much more difficult than keeping them from doing something destructive. Induced action must overcome inertia. Blocked action simply maintains inertia.

Resource restoration has occurred in many settings—from the Sahel to the Himalayas (Timberlake, 1987; Aga Khan Foundation, 1988). Terraces, catchment basins, hedgerows, and tree planting are examples of small-scale restorations. The motivating factor is usually despair or a sense of loss. Perceived degradation, when combined with community organization and skilled advice, can lead to improved infrastructure and vegetation. And efforts often begin with a vision of a better place, but little in the way of clear targets. Indeed, empowerment of the community can be as important an objective as any immediate physical result.

On a large scale, government afforestation programs illustrate restoration. They usually result when policy makers are convinced of the seriousness of the degradation problem. Aerial photographs showing how political and vegetative boundaries coincide, as in Lesotho, or visits to the countryside showing how mud slides result from lack of tree cover, as in Nepal, can be persuasive tools for convincing them.

Restoration could also take the form of a policy to create a substitute for non-renewable resources. For example, Herman Daly has suggested that a tax on coal mining could be used to establish a fuel wood plantation that would have a sustainable annual yield of British Thermal Units (BTUs) equal to the BTUs contained in the ore extracted annually from the mine (Griesinger Films, 1991). If such a practice were to emerge, the initial mobilization and short time frame of most restoration projects would evolve into a long term management activity.

Three elements separate restorative from prohibitive approaches. First, restoration is based on positive action that needs organizational and promotional skills, whereas prohibition requires investigative and enforcement skills. Second, the motivations for the two activities are quite distinct. Obtaining benefits is not the same as avoiding burdens. And third, the view of people is radically different. When prohibiting actions is the objective, people are the problem. When restoring resources is the objective, they are the solution. Negative reinforcement is the foundation for prohibition, while a positive approach to reinforcement is the basis for restoration.

This does not mean that coercion is never part of restoration. There are examples of the use of prison labor to rebuild bridges and cattle herders have been forced at gunpoint to construct and use cattle dipping facilities. But the success stories are imbued with positive leadership and an operating style that is more promotional than coercive.

Promoting New Behavior

Restoration is a one-time effort requiring limited energy and resources. No fundamental change in behavior is necessary. There may be some maintenance of a new structure, but the majority of the effort is a one-time affair.

A very different thing is transforming routine behavior—getting people to do something they have never done before. Recycling comes to mind in the developed nations. But this is not something that needs to be encouraged in the third world. It has been a natural outgrowth of poverty—very little is left unused in the poor nations, offering testament to the power of context.

Supporting new behavior is also very different from stopping old behavior. In fact, mixing the two functions seldom works. A case in point involves the satellite organizations established as part of rural development projects. These organizations were often conduits for credit funds. The idea was that group lending would fit local cultural practices and would substitute group social pressure for collateral as a means of ensuring repayment. But then the organizations were expected to expand their role to encompass creative entrepreneurial undertakings and political representation. Of course, they failed to make the transition (Honadle and Vansant, 1985). Likewise, giving agricultural extension agents dual functions of promoting new crop husbandry practices and collecting credit payments causes both functions to suffer (Honadle, 1982 b). Mixing control and catalyst functions does not work.

In the natural resources realm, social forestry programs are one of the more common initiatives to promote new human behavior. Planting and caring for trees, incorporating trees and shrubs into the agricultural landscape through inter-cropping, living fences, or planting for terrace support are often found as components of social forestry programs (Gregersen and others, 1989). Individual and community wood lots, fuel wood price raising, charcoal production and public nurseries also often appear as program components. Indeed, such programs tend to be complex, reflecting the nature of behavioral innovation.

Another common type of effort to promote new behavior is range management policies. This is an area where the experience has been largely disappointing (Sandford, 1983), although attempts to mimic natural systems have registered some success.[2] Changing ingrained practices is extremely difficult to do because it involves altering a web of social, political, economic, and sometimes even religious elements. And, in Africa, many attempts to change range management practices occurred during periods of drought, when the systems were under other stresses, when the tendency was to use traditional survival algorithms to last out the stress period, and before imported approaches could demonstrate whether they would even work.

The greatest experience base in the third world, however, is probably that found in the agricultural extension history. This history suggests that success requires a combination of sound technologies, locally appropriate education and demonstrations of new behavior, and support systems (including input supply and marketing channels). Weakness in any of the requisites threatens performance.

Such lessons should not be lost on environmental policy makers. Developing new behavior will require investment in delivery and support systems that promote and maintain new ways of doing things. Indeed, it even may require help in overcoming the risks involved in a transition from doing things one way to doing them another way.

For example, the beginning of this chapter mentioned the Ecuadorean homesteader cutting trees to establish title to land and the Filipino fisherman destroying the coral reef to meet today's need for subsistence. Changing land title policy may involve little beyond the stroke of a pen, attendant publicity, and a new boundary marking system. But protecting the reef may involve alternative employment opportunities for the fisherman combined with protective measures. Or, the development of reef-neutral harvesting methods that are efficient and less costly than destructive approaches would be another way of bringing about behavior change. Both are far more complicated matters than simple protection, which is not likely to work.

Preserving a Resource

Wildlife preservation is often the motivator behind this approach, whether it is the call to save the Serengeti ecosystem in the 1960s, or the cry for protection of biodiversity in the 1990s. Whether it is the lemurs of Madagascar or the rainforests of the Amazon basin, the automatic preservationist approach is to create reserves where human access is severely limited and sometimes completely prohibited.

It may involve positive action, such as passing laws setting up parks and preserves or moving people out of wildlife areas. But it invariably involves inaction, such as *not* building roads into selected areas. Indeed, policy intentionally emphasizes not helping people to move toward the area in question.

Proximity of people makes this objective harder to achieve. Protecting a remote island in the Philippine archipelago or the hard to reach Nyika Plateau in Malawi is far easier than sequestering Khao Yai National Park, which is just a few hours drive from Bangkok, or Yosemite National Park, which lies near the concentration of people in California. There is also much less resistance to the policy if no one is already living within the intended boundaries of the protected area. Sometimes limited numbers of people are allowed to continue to live within the park boundaries with restrictions on their activities and movement, such as with the Maasai herders in Tsavo park in Kenya. But resource preservation generally aims to separate people and natural areas.

Indeed, this is often seen as "anti-people" by local populations. The ecological reasoning of outside "experts" is not well understood and local populations do not accept it as a legitimate decision. They see it as theft of their livelihood, theft of their birthright, and destruction of their continuity of place, and they respond accordingly.

In fact, the case has been made that the separation of people and wild lands in the developing world, especially in Africa, has been based on the ignorance of outsiders about the impact of local societies on the environment (Bonner, 1993). The race by outsiders to save Africa's animals has often involved the imposition of external views in the absence of knowledge about the relationships between indigenous societies and their environments (Grzimek, 1970; Hayes, 1977; Holman, 1967). A sequence of first separating indigenous people from wildlife and then injecting the presence of tourists into the same locale often leads to greater physical decline than would have occurred by managing the blend of local human livelihoods and indigenous flora and fauna. Indeed, the real cost of foreign exchange produced by tourism can include both political conflict and ecological decay.

Pressure to preserve an area tends to incite and mobilize those intent upon mining the riches from that area. In Cameroon, for example, fears of import restrictions in the markets of the developed world propelled timber companies to plunder the rainforest at an even faster pace (Horta, 1991). The time frame is compressed as exploiters perceive a need to get it all now or forever lose it.

This consequence has much in common with the first one—the decline in damaging behavior. But it is also aimed at the avoidance of future damaging action. It may not be involved with stopping something that is already happening. To the extent that it is future-focused, it is an easier task. Avoiding opportunities for temptation is less wrenching than unlearning routine behavior.

Increasing Efficiency in Using a Resource

Increasing efficiency is generally a technology-driven process. It may be high-tech, such as improved steel-making furnaces in Europe, or it may be low-tech, such as new household stoves in Africa. Each minimizes waste in the use of a resource and thus stretches the longevity of resource stocks or lowers the cost of protecting other resource stocks.

Promoting increased efficiency is directed less at individuals than the other targeted consequences noted above. The target for this behavior change is more likely to be a household or corporation than an individual. Corporate activities especially are apt to be scrutinized to find opportunities to increase efficiency.

Efficiency may be mandated, but without alternative practices or technologies such a mandate has little meaning. Thus, a system for identifying, producing, distributing, and servicing more efficient technologies is a prerequisite for more efficient resource consumption. Research and development (R&D) are integral to improved efficiency. A policy commitment to this objective may involve a program of research and testing supported by public funds. In some cases no hardware, just new ways of organizing human effort, are involved. But the more common case requires new equipment of some type.

Demand management approaches to electricity use illustrate this—educating people to use equipment during off-peak hours, turning off lights not needed, turning down the temperatures of water heaters and pumping less water are behavioral, but installing flourescent bulbs or improving insulation both require the availability of improved technologies. And technology can even be substituted for behavior—computerized thermostats, timers, and pumps all change system performance more reliably than trust in human action. When equipment can be imported or when it is already available locally, the cost of R&D will be less, but even here adaptation is usually needed. In any case, the implications of this behavior change are very different from those introduced in the preceding sections above.

These five targeted consequences allow the specification of who the policy will affect. But there are also objectives where the targets are not nearly so clear.

SYSTEMIC CONSEQUENCES

Economic development is often contrasted with economic growth by emphasizing the importance of structural transformation of an economy. Growth can mean just more of the same things, whereas development means new things, such as the increased contributions of services and industry to Gross National Product (GNP) and the decreasing importance of agriculture.

Likewise, the transition to a "green economy" involves changes of a systemic nature. The intended consequences of environmental policies may involve bringing about such transformations as creating markets for environmentally benign substitutes for destructive materials or processes, increasing the performance capacities of organizational networks, or altering basic decision processes within households, communities, or public or private organizations.

Just as an environmental problem may be caused by non-point sources, so too the solutions to a diffuse set of interactive problem causes will be non-point solutions. This is the essence of systemic strategies—redesigning sets of interactions rather than zeroing in on one specific behavior or actor.

Creating Markets for Substitute Materials/Processes

The creation of markets evokes an image of a simple act—the signing of an accord to establish a stock exchange or the stroke of a pen changing government procurement policies. Each establishes new avenues for investing assets or supplying goods and services and therefore brings a market into being. It makes market creation look easy.

Indeed, to a neo-classical economist, nothing is more natural than a market. The eruption of shadow markets in the wake of the introduction of planned economies in the third world provided thirty years of evidence that this was the case.

But the collapse of the Soviet empire jolted this accepted knowledge. Attempts to reintroduce market thinking and market behavior in a place that suppressed markets for nearly four human generations is showing very clearly that markets exist within a web of supporting institutions. Without such things as contract law, property rights, bankruptcy law, processes for pursuing tort claims and enforcing contracts, and tolerance for entrepreneurial activity and profit margins, markets do not thrive. For markets to work effectively, neither the legal tapestry nor the social fabric can stand in overt opposition to market dynamics.

This suggests that policies to create markets for environmentally benign materials and processes will sometimes involve systemic changes. Changes of relative prices may work in market-familiar settings, but other places will need drastic reconstruction of institutions. Such places as Zambia, Mongolia, Tanzania, and the nations of eastern Europe come to mind most readily as candidates for reconstruction.

This is important due to the present eagerness to use market mechanisms to cure environmental ills (Anderson and Leal, 1991). The perceived efficiency and administrative simplicity of markets is elevating them to a preferred status among alternative approaches (World Bank, 1992a). But the ease of introducing market processes may be overestimated because the mental picture of a "stock or commodities trading market" dominates much thinking, and because discussions of prior third world experience with water markets often understate the extended time required to make them work (Maas and Anderson, 1976).

Trading water rights, pollution credits, or land rights needs a support structure to work. The free information flows associated with laissez faire operations will need bolstering by green labeling laws (OECD, 1991b), measurement standards and monitoring structures and operations. Market-based environmental policies will encounter a myriad of complementary actions without which they will not work. The consequences will be systemic rather than limited to the behavior of a few folk.

A widespread panacea for protecting natural areas is eco-tourism (Edington and Edington, 1986; Whelan, 1991; Lindberg, 1991). This is a strategy of substituting income from tourists who pay to see a protected area or species for income from actors who consume or destroy the species or landscape. This strategy is characterized by a combination of command and control to stop the consumptive behavior and marketing to promote the benign behavior.

Similarly, extractive reserves offer hope for the use of non-timber woodland products to generate income without depleting old-growth forest areas (Schwartzman, 1992; Salafsky, Dugelby and Terborgh, 1993). But it is becoming increasingly obvious that securing markets for substitute materials and processes, and then supplying the substitutes, can be a very complex endeavor (Chambers and Leach, 1989; Kiss, 1990; Redford and Stearman, 1993). In fact, many relationships within society may need to be realigned and many new organizations may need to be established for a market-based approach to work.

One theoretical assumption of neo-classical economics is perfect information. Buyers and sellers, producers and distributors all need to be informed. This requires communication infrastructure and open, or at least obtainable, access to information. This is a systemic prerequisite for market efficiency.

Developing Adaptive Capacity and New Decision Processes

Installing markets involves getting people to respond to new cues. Installing adaptive capacity involves creating the ability to continue to assess a changing environment and to construct new responses to the changes.

This sounds easier than it is. Simple education and training of individuals does not always produce the desired results. Visionary leadership, organizational incentives to embark upon untested activities, an experimental view of the world, risk-taking, and the desire to innovate are hard to induce.

Adaptive capacity may involve greater legitimacy for heretofore marginal members of society. Or, it may mean a direct threat to traditional leaders and the basis of their power. In either case, the task is tough and resistance can be expected.

Human resource development is a central aspect of capacity building, but simultaneous stimulation of markets and control of deleterious behavior may be needed. Although a core element in the creation of green societies, economies, and institutions is consciousness raising, much more is involved. Basing actions primarily on ecological impact requires inserting a new set of criteria into routine decisions that previously responded solely to an economic marketplace, a kinship calculus, or a political pressure group. This was stated simply and directly by Schmidheiny in preparation for the United Nations Conference on Environment and Development (UNCED) held in Rio de Janeiro, Brazil, in 1992. He said:

Sustainable development will obviously require more than pollution prevention and tinkering with environmental regulations. Given that ordinary people—consumers, business people, farmers—are the real day-to-day environmental decision makers, it requires political and economic systems based on the effective participation of all members of society in decision making. It requires that environmental considerations become a part of the decision-making processes of all government agencies, all business enterprises, and in fact of all people. (Schmidheiny, 1992: 7)

The assumption here is that rational people will respond to the need to survive, and that the need to survive will dictate new ways of judging the quality of decisions and processes. But understanding the consequences of alternative decisions will require education. And for such education to be well-grounded, new research into product life-cycles, long-linked chains of cause and effect, and the environmental tradeoffs among various ways of doing things will be necessary.

But knowledge alone may not lead to the desired results (James and Gutkind, 1985). Unless cultural norms and role models support new behaviors and unless social rewards accrue to those who adopt new "green" approaches, the legions of adopters may be meager indeed.

In a firm or organization, strong leadership may be able to praise the virtuous, reward the innovators and penalize the laggards, but in large scale societies the effects of personality have been far less than was once anticipated. Creating new decision processes is not easy, nor is it necessarily a direct process.

Solving social problems will be integral to creating new decision processes. In inner city America, for example, solving the drug problem involves creating social mobility through avenues separate from the drug trade. Likewise, in some parts of the third world solving land tenure insecurity may need to precede green legal codes and market-based policies for the outcome to be positive (Panayotou, 1993). In other situations, secure tenure can facilitate the transfer of the land to ravagers (interview with James Nations, 1993). The question is "what social conditions cause inappropriate or destructive behavior?" (Friedmann and Rangan, 1993) This is the key to developing adaptive capacity and redesigning decision processes.

Replacing Accounting and Measurement Systems

Adaptive capacity requires research and monitoring capacity, as well. Systemic development will be information intensive and it will involve the creation of new measures of success. For example, much of today's frustration with the inability of GNP to capture changes in wealth and well-being reflect the inadequacy of the measure. Economic activity that generates employment through the destruction of natural resources is treated as a positive contribution to national wealth equal to an activity without destructive impact that generates the same amount of income. Indeed, a leading economic indicator is new housing starts, which also suggests loss of open space, habitat destruction and conversion of land to new uses. It is not sensible for this to be treated the same as the same amount of income generated from ecological restoration. The measurements that we use to assess economic strength clearly are primitive and misleading. In fact, the faster we deplete our natural resource stocks, the stronger the economic indicators tell us we are.

Alternative indices are promoted by such organizations as the United Nations Development Program (UNDP) and the World Resources Institute (WRI). The UNDP has developed a "human development report" which has measures other than strictly economic ones to assess national well-being. For example, one indicator is the ratio between soldiers and teachers in a country. The annual report acts as a shadow account that balances the measurement of the World Bank's annual *World Development Report*. But this means that the UNDP analysis is simply an addition to the traditional economic review of the Bank. It is not a substitute for GNP. It merely adds a different perspective.

The World Resources Institute argues that the economic accounts used today are faulty and should be replaced. A more accurate measure is needed to show actual economic strength and weakness.[3] But it is much more difficult to substitute a new measuring system than it is to just add another report series to the annual output of stock-taking studies. Indeed, the adoption of a totally new system of national accounts would need to be reflected in organizational accounts as well—both economists and Certified Public Accountants would need to relearn their trades.

The effort to make our accounting systems reflect the real costs and benefits of goods and services is ongoing under at least three different labels. One is "natural resource accounting" or "valuation." This attempts to attach values to natural resource stocks and to the services that we receive from them, services such as shade, reduced wind speed, air and water filtration, lowered topsoil erosion, biological control of pests, moisture retention, toxin suspension, gene pool robustness, aesthetic enhancement, atmospheric temperature moderation, medicinal applications, and many others. The purpose of this is to allow us to figure out the cost to us of what we are losing.

Another label is "full cost accounting." This is an attempt to assess the total costs, not just financial costs, of human action. For example, the installation of an electric transmission line would need to include the dollar value of aesthetic disruption on a tourist economy, the damage to biological communities, the impact on invasions of exotic species, and many other changes, in its calculation of costs—not just the financial cost of construction.

The third label is "life cycle analysis" which ventures well upstream and well downstream from a product to incorporate the pollution and other costs of production, use and disposal into its price. Producing an automobile, for example, would include in its cost the environmental damage resulting from mining the ore to make it, extracting, refining and delivering its fuel, operating it (such as airborne emissions, salt on roadways, paving land, oil seepage) and even land use and contamination issues attendant to disposing of any of its parts that are not directly recycled. An emerging perspective within life cycle analysis incorporates precautionary, preventive, democratic, and holistic principles, questions the very need for products, and uses the term "clean production" to integrate these dimensions. (Thorpe, 1999)

All three of these approaches converge on the point where the economy meets nature. And none of them are peripheral to the way a society does its business. The adoption of such departures will permeate the entire culture. They will change the ways that bankers, insurance companies, government agencies, merchants, industries and households conduct their business and assess their performance. They are systemic changes requiring reflective, critical and substitutional processes that go to the core of social interactions — they are not simple actions resulting from a single policy change.

Thus, the discussion above supports the idea that we should know what the objectives of a policy are—not just whether the objectives are clearly stated. In fact, some objectives may not need extreme clarity, and some might even benefit from fuzziness in the short term. But if we don't know what we are trying to do, we are not likely to be able to figure out how to do it.

Figure 2–A identifies some of the different characteristics inherent to attempts to achieve different behavioral objectives. This figure displays and clarifies the variety of implementation strategies that support the achievement of different consequences.

As both the discussion and figure above suggest, different intended consequences will involve very different implementation strategies. Without understanding this, it is unlikely that policy promoters will choose effectively among the clusters of policy options or the implementation mechanisms introduced in the next chapter.

NOTES

1. The authors of this report do not bear full responsibility for this shortcoming. They were simply responding to a scope of work for contracted research. So those writing the terms of reference bear some responsibility. But the responsibility goes further than this—professionals in the field of development management have focused too intensely on the processes and attributes of management mechanisms and not enough on context and on the fitting of attributes to context. And even after a well-received study suggested that contextual factors may be important in determining the appropriateness of western management methods (Kiggundu, Jorgensen and Hafsi, 1983), the focus failed to change. Thus, as described in Chapter One, the problem of looking under the lamppost remains.

2. The exception to the pattern of generally poor results with range management projects is truly exceptional. Instead of lowering cattle numbers to fit "carrying capacity," cattle numbers were actually increased, but active management (short-term rotation from area to area) was used to mimic the conditions that characterize wild animal movement. Quick rotation imitated ungulate behavior in the face of active predation. This gave grasses the time to prosper, improved ground cover, and led to both higher herd levels and better range conditions. And it probably improved the nutritional intake of the cattle as well. Countries where this was done include Zimbabwe and the United States. (Savory, 1988) But even this experience has generated doubts about its effectiveness.

3. Herman Daly argues that we should move toward the tallying of a sustainable net product (SNP) consisting of three accounts—a benefit account, a cost account, and a capital stock account. See Daly, 1996.

FIGURE 2–A: Characteristics of Different Behavioral Objectives

INTENDED BEHAVIORAL CONSEQUENCES	CHARACTERISTICS OF THE MEANS TO ACHIEVE THE CONSEQUENCES
less damaging activity	1 - often control-based, using disincentives 2 - needs clear objectives 3 - needs clear link to costs or punishment associated with damage 4 - can focus on individual behavior
restoring resource	1 - positive action must be mobilized using appropriate social units and processes 2 - vision needed but clear objectives not necessary 3 - high cost, often last resort 4 - short time-frame
new conserving behavior	1 - extended time-frame 2 - routine, non-dramatic focus 3 - incentives important 4 - programs often complex 5 - can be individual or group based
sequestering resource	1 - resource should be site - bound 2 - can involve inaction 3 - remoteness helps 4 - anti-people image 5 - focus on resource, not people 6 - extended time-frame 7 - scale and infrastructure important
more efficient consumption	1 - technology key 2 - social unit focus 3 - incentives important 4 - extended time-frame
substitute market creation	1 - supporting institutions needed (consumer education, trading mechanism, policing, production, legal rights) 2 - information flows important 3 - may combine positive and negative action 4 - not necessarily targeted; system-wide focus
new decision processes/adaptive capacity	1 - knowledge and education key 2 - needs support for innovation and risk-taking 3 - research and development focus 4 - strong leadership required 5 - not targeted; system - wide focus 6 - objectives fluid and broad
new accounting / measurement system	1 - legal consequences for non-shift 2 - must be system-wide, not piecemeal 3 - requires prior development of new system 4 - strong leadership required 5 - must be preceded by strong educational campaign

3

Twisting Pathways: Alternative Implementation Strategies

Word association games can tell us much about the blinders we wear as we engage the world around us. When the word "policy" is tossed out, the response is often "enforcement." When "implementing . . ." is presented, the response is often one of the trailing word "agency." But both reactions indicate a case of tunnel vision—we see only a limited range of alternatives compared with what is possible.

The way that our conceptual categories blind us to reality was captured by Robert Pirsig. Using motorcycle maintenance as a metaphor for life and the quest for understanding and completeness, he wrote:

. . . there is a knife moving here. A very deadly one; an intellectual scalpel so swift and so sharp you sometimes don't see it moving. You get the illusion that all those parts are just there and are being named as they exist. But they can be named quite differently and organized quite differently depending on how the knife moves.

For example, the feedback mechanism which includes the camshaft and cam chain and tappets and distributor exists only because of an unusual cut of this analytic knife. If you were to go to a motorcycle parts department and ask them for a feedback assembly they wouldn't know what the hell you were talking about. They don't split it up that way. No two manufacturers ever split it up quite the same way and every mechanic is familiar with the problem of the part you can't buy because you can't find it because the manufacturer considers it a part of something else.

It is important to see this knife for what it is and not to be fooled into thinking that motorcycles or anything else are the way they are just because the knife happened to cut it up that way. It is important to concentrate on the knife itself. (Pirsig, 1974: 79)

Pirsig's warning is especially pertinent to the issue of environmental policy today. Many of our debates and negotiations are little more than artifacts of the analytical tools we bring to the table. Economists do not grasp or value the points of environmentalists; technicians do not appreciate the needs of politicians; and natural scientists fail to see the arguments of social scientists because each has analytical categories that ignore (and perhaps even deny) the assumptions and objectives of the other. And all of the above may ignore the needs, knowledge, and risks that characterize the daily lives of people who live near to, and may rely upon, the resource whose fate is so vehemently debated.

This chapter introduces a policy and policy implementation perspective based on the social sciences. The purpose is to present a scalpel that cuts through our blinders and exposes new insights.

THE POLICY PARADIGM

We are concerned with public policies—the stated objectives, operating rules, laws and legal codes, and binding agreements that are endorsed, pronounced, and promulgated by governments and international organizations. Such policies emerge from three distinct end-of-process actions throughout the world. They are:

- *legislation*: parliaments, legislatures, and councils of leaders propose or pass laws that become the rules of the land;

- *executive decrees*: presidents, prime ministers, governors, or executive agency heads produce procedures, rules, statements, policy papers, guidelines, and objectives that others must adhere to; and

- *contracts and agreements*: governments enter into loan agreements, project agreements, pacts, conventions, accords, protocols, treaties, and memoranda of a contractual nature with other governments, international organizations, corporations, and non-governmental institutions.

But how observers view these policies depends upon their backgrounds. Policy analysts tend to be technicians—people trained in economic and planning tools and techniques and people susceptible to the belief that formal adoption of an approach means that it will be followed. From this perspective, policy is an independent variable that explains consequences, and once the policy is adopted the job is done (Allison, 1975). Implementation is someone else's concern.

Institutional specialists, however, take a different view. First, in the tradition of political scientists, they see policy as an outcome of a political process—a result, not a cause (Dye, 1972). Second, in the tradition of organizational sociologists, they distinguish between policy that is *espoused* and policy that is *in use* (Argyris and Schon, 1976).

The *espoused policy* may result from pressure exerted by international donors or conservation organizations and it may not reflect the true intentions of a national leadership. Indeed, it may be designed to mask the actual objectives, which are revealed by the policy in use. For example, declaring a ban on the sale of ivory can both promote a clean image to the tourist industry and create a shadow market more easily monopolized by key officials. Thus, it can be an unreliable guide to human behavior. In fact, espoused policy may be no more than a smoke screen intended to give hope and false support to the proponents of a low priority policy (B. Honadle, 1993).

But the *policy in use* does have predictive power. It reveals how people are likely to respond to new initiatives. It goes beyond formal, open rhetoric and gets at cloaked behavior. From the institutional perspective, formal policy pronouncements are less important than the actual behavior that follows (Honadle, 1982a). Policies are not single-time vaccinations that eradicate plagues. Instead, they are merely statements of intention that need constant support. In fact, the wording of the policy will imply the types of support needed. Policies may be worded three ways.

- The first approach is to use wording that is *directive*. That is, someone "must" or "will" do something. To work, such a command needs control mechanisms built in.

- The second approach is to use wording that is *restrictive*. That is, someone "must not" or "will not" do something. This too, requires control mechanisms.

- The third approach is to use wording that is *enabling*. That is, someone "may" do something. This relies more on incentives and human initiative than on control.

But with any of these three constructions there is another aspect of the policy paradigm that is key—the view of policies as *hypotheses* about how one factor influences another. No matter whether directive, restrictive or enabling approaches are used, the link between action and result is fraught with uncertainty. Due to our existence in a world of probabilities, rather than certainties, policies are hypotheses. We do not know for sure what the consequences will be when a new policy is set forth. Policy change is an experiment. And experiments require testing, not just imposition.

This generates a need for flexibility, social learning and adaptation in policy implementation (Sweet and Weisel, 1979; Korten, 1980; Rondinelli, 1983; Honadle and VanSant, 1985; White, 1987; Honadle, Grosse and Phumpiu, 1994). Indeed, the implementation of a policy is the test of it—it is an experiment.

From the above, it is apparent that the source, construction and hypothetical nature of policies influence implementation requirements. But if we are to further our understanding of the relationship between policies and environmental conditions, we need to go beyond general characteristics of policies. We need to find a framework that helps us see alternative categories of policies for influencing natural resource stocks and promoting sustainable economic activity.

A PERSPECTIVE ON POLICY OPTIONS

Natural resource policies do not aim at natural resources—they aim at human behavior. Nature does not respond to human policy. And yet typical policy categories are based on the resource base involved—water, forests, wildlife, etc. or else they reflect economic subject areas such as trade, tax, or monetary policy.

But both of these types of categories are too narrow—they impose blinders that perpetuate single-discipline prescriptions and avoid alternative approaches to changing human behavior. What is needed is a set of policy types that goes beyond these limitations. One way to depict policy options would be to compare and contrast the amount of control that is retained by the people who use a natural resource. When we do this we can see three basic clusters of policy options. They are:

- command and control;
- direct incentives; and
- stakeholder self-management.

These options are arranged above in an ascending order of control retained by resource users—command and control allows little autonomy, stakeholder self-management allows near total autonomy, and incentives fall between the two extremes because they provide penalties or rewards for different actions but leave the decision up to the user. Each cluster is elaborated below.

Command and Control

The first policy cluster is *command and control*. This is familiar to everyone. It is the sequence of decree or legislation followed by regulation. It is a common approach to protection of wild lands, species, and resources. National environmental protection acts, enforcement of maximum pollution levels, setting aside of natural areas for parks or research, limiting timber cutting or requiring certain silviculture practices, restricting access to precious minerals, banning commerce in ivory or endangered species, or even forcing compliance with hunting and fishing laws all provide examples of this approach. It is perhaps the oldest natural resource policy practice, dating back thousands of years to the protection of royal game.

The image it evokes is one of denying people the right to do something and it explains why natural resource management agencies and regulatory bodies are seldom popular. Indeed, command and control policies are directive or restrictive in nature. China's population control program is an example of this approach to limiting human population growth and mitigating its effects on the environment.

One of the major consequences of command and control approaches in third world settings has been the formation of shadow markets. To professional policy analysts this is an unintended effect. But to many local political actors it appears as a major benefit because it allows the charging of bribes, it concentrates control, and it creates opportunities for amassing wealth. Corruption flourishes under command and control (Ledec, 1985; Gallagher, 1991).

Self-management

The second option is the polar opposite of command and control. It is *stakeholder self-management*. This is giving total control of a natural resource to a local population and trusting that they will manage it on a sustainable basis. The assumption is that they have a long-run stake in maintaining the resource and the most effective way to ensure sustainable use is to let them do it.

Self-managed situations result from enabling policies that devolve authority, recognize and legitimize pre-existing traditional authority structures and resource management systems, and foster local empowerment. Decentralized administrative strategies, privatization initiatives, Aborigine reserves, community-based natural resource management programs (CBNRM), local eco-development programs and many social forestry efforts throughout the third world provide variations on this theme (Gradwohl and Greenberg, 1988; Poffenberger, 1990; Korten, 1986; Hanke, 1987; Gregérsen and others, 1989; Silverman, 1990; Institute for Rural Development, 1991; Hitchcock, 1997; McCormick and Honadle, 1999). In the area of population planning this is the most common approach—people are supplied with information and contraceptives, but they are responsible for what ensues.

Stakeholder self-management is increasingly appearing as a conservation and management strategy throughout the globe, from the Philippines (Christie, White and Buhat, 1994) to the United States (Fortmann, 1994). And the strategy crosses sectors—coral reef management, forest management and wildlife management, to name a few, reflect this approach. In the wildlife area, for example, giving local communities a monopoly on the meat, hides and other products from wild animals, creates an incentive for sustainable management. An oft-used illustration is Zimbabwe's CAMPFIRE[1] program, where local communities determine what level of hunting is appropriate and they receive income and employment from wildlife-viewing tourism.

Self-management is an option that is surrounded by mixed reviews. It has the advantage of being administratively efficient by not requiring a great superstructure of directive machinery. And it has the appeal of allowing autonomy to regional groups and distant populations. But it also carries the fear that resource use will follow the path to the "Tragedy of the Commons" and lead to irreparable resource depletion as each individual maximizes his or her immediate gain (Bromley, 1992), it demands local—sub-national—management capacity (Bromley and Cernea, 1989), and it thwarts those seeking to impose a regime of bribery and tribute. So, it offers hope but it also raises fears. Indeed, one observer of common property management problems has noted that neither command and regulation nor private self-management provide adequate solutions to the problem of resource depletion (Ostrom, 1987). Broader perspectives are needed.

Self-management also requires a trusting environment. That is, neither experts who conceal their paternalistic attitudes through the trappings of scientific knowledge nor national politicians who conceal their disdain and suspicion of regional groups are likely to promote stakeholder self-management. When environmental protection masks power plays, this option is seldom endorsed. But the third cluster, direct incentives, can be found in a wide range of settings.

Direct Incentives

The third policy cluster, lying between command and control and self-management, is *direct incentives*. Discussion of pollution credits, water markets, stumpage fees, green labeling, and other examples of market-based environmentalism fit into this category. (Anderson and Leal, 1991; Armitage and Schramm, 1989; Maass and Anderson, 1978; OECD, 1991b) This focuses on the structure of benefits and costs surrounding a particular resource, product, or process. Direct incentives enable behavior, or they require it as a qualification for rewards. In Ecuador, for example, pioneers moving into rainforest areas are required to clear the land to establish a legal claim to it. This offers direct rewards for environmentally destructive behavior.

Subsidized costs for logging equipment, agriculture or aquaculture inputs, marketing assistance or the establishment of favorable prices for certain products are other examples of direct incentives that are commonly found in third world settings. In fact, attempts to establish direct incentives have been integral to development projects for decades and both bilateral and multilateral donors routinely incorporate price incentives into project investment packages.

Although subsidies have bounced in and out of favor, producer responsiveness to farm gate prices has been raised to nearly mythical proportions and direct incentives are likely to occupy a key role in donor strategies for some time to come. This is especially true given the present enthusiasm for using markets to solve environmental ills.

This typology helps us to see the differences among various policy options for protecting natural resources. But something is missing. Policies that are not intended to have any effect on natural resources may be very important. In fact, policies that are seemingly unrelated to environmental issues may spill over into the natural resources area and negate the effects of forestry, wildlife, or pollution abatement policies. The fourth policy cluster, then, is:

- indirect incentives.

Adding this option completes the typology of clusters of environmental policy initiatives. But, even more than the three above, it too needs to be elaborated and explained.

Indirect Incentives

The fourth cluster is *indirect incentives*. This is a particularly pervasive set of influences on resource use because it usually results from the pursuit of other, seemingly unrelated, objectives and thus it is ignored by most environmental observers (exceptions are Brown and others, 1991; and Von Weizsacker and Jesinghaus, 1992). Indeed, this policy cluster needs more detailed treatment here because it is less understood and less likely to be examined when policy options are discussed.

Policies promoting labor-intensive technologies often have benign environmental impacts due to limitations on the use of large, environmentally destructive equipment. Tax codes intending to promote reinvestment in private enterprise, structural adjustment programs opening up previously closed economies, or tariffs biased against specific technologies can be examples of this policy cluster. Although they may encourage either protective or exploitative behavior, indirect incentives often work against sustainable resource use. But because of their indirectness, the impact is seldom understood.

In the United States, for instance, carbon-based pollution may result partly from public and private employer policies that prohibit commuting by telephone and thus increase automobile use. Wetland loss in the upper midwest may result partly from a tax code that allows rapid depreciation of farm machinery (thus leading to the purchase of larger equipment and the chopping of hedgerows and filling of wetlands to minimize the need to turn the equipment around during operation). Tax codes in some third world nations may also encourage inefficient or environmentally destructive behavior. Across-the-board export levies, for example, can make non-timber forest products less competitive for producer attention and speed deforestation.

Even procurement policies of governments or international organizations could be redesigned to require environmentally sound processes for producing goods and delivering services. This is an often ignored area with great potential. For example, in 1992 the World Bank procured 8.5 billion dollars worth of goods and services worldwide, but nowhere in its sample bidding documents does it suggest that environmentally sound production processes will influence its choice of supplier or that borrowers should take such considerations into account (World Bank, 1992b; World Bank and Inter-American Development Bank, 1986). Indeed, when such indirect and operational policies are inconsistent with espoused ones, they can sabotage the more direct and publicized policies. Spillover from policies in other sectors can contaminate the environmental sector.

Foreign aid operations can indirectly exacerbate resource depletion. For example, a program of financial support to the government of Chad would have cushioned the impact of a change in tariff levels until it was discovered that there were environmental consequences. Lower import duties on automobile and truck parts could have speeded migration to marginal areas by improving the national transportation fleet. The environmental impact of a macro-economic policy reform might not have been contained. The possibility of this led to a donor supporting other policy reform instead (E. Karch interview). The spillover from one sector to another was not immediately obvious, but later it was determined to be highly probable.

The proliferation of macro-economic adjustment efforts in the third world also has created many opportunities for the inadvertent imposition of indirect incentives for environmental destruction. Zambia in the mid-1980s provides an example.

The introduction of a new system for setting the foreign exchange rate of the Zambian Kwacha resulted in a setback for investment in agricultural and natural resource initiatives. The focus was on replacing an administered foreign exchange system with a market-based one. The solution to poor economic performance was cast as reforming economic policy and introducing a market regime. No one anticipated any impact on the environment.

The way the new system worked was as follows: Actors interested in purchasing foreign currency held by the central government presented bids to a bank on a certain date each month; The bids indicated how much currency was desired and how many Kwacha would be paid; The bids were examined, the high bidders received their allocation and the next highest received their amounts until the total consumed all the foreign exchange that the government held at that time; The high bid established the official exchange rate for that month. Some modifications occurred during the life of the auction process, but the changes made no difference in the impact on agricultural and natural resource investment.

An examination of the successful and unsuccessful bids revealed a disturbing pattern. The winning bids invariably were for activities with fast turnarounds that were often simple middleman operations importing consumer goods and equipment. Lower risk, rapid cost recovery activities allowed higher bids. But higher risk, longer turnaround activities (such as forestry, fisheries, and agricultural investments) forced lower bids. This was reinforced by subsidized food for urban populations, which made agricultural investment even less profitable. Thus, implementing the currency auction process in this setting exacerbated a pre-existing bias against natural resource investment and it provided an indirect incentive for further sectoral imbalances in a country characterized by major environmental deterioration and chronic food short-ages.

This case shows how indirect incentives can be both important and difficult to identify. Indeed, interactions among multiple indirect reward systems can weave a web of barriers to sustainable resource use. Natural resource policies may be ineffective because key policies may lie far outside the commonly defined natural resources sector. When such policies are highly insulated from affecting human-natural resource interactions, then they are of no concern to environmental policy professionals. But when the insulation is low and their impact crosses sectoral boundaries, then they may be major determinants of environmental quality. (Also see: Runge, 1997)

The fourth cluster, indirect incentives, completes our typology of policies affecting natural resources. Although parts of this typology have been used by others (Eskeland and Jimenez, 1992), Hans Gregersen of the University of Minnesota was the first to suggest the four broad policy categories as a viable alternative to traditional economic policy sets such as fiscal policy, trade policy, etc., and as a substitute for sectoral policy divisions such as population policy, macro-economic policy, transportation policy, agricultural policy, and so forth (Gregersen and others, 1991).

The value of this typology is fivefold: first, it is comprehensive enough to include nearly any conceivable policy; second, it allows a direct focus on environmental consequences whether they are intended or not; third, it provides an integrated and applied focus rather than a set of categories based on disciplinary or institutional slices of reality; fourth, it allows the formulation of hypotheses about the sustainability of some policies; and, fifth, it is compatible with a behavioral focus.

Figure 3–A contrasts the first three clusters along a vertical axis representing autonomy and places the fourth cluster on a horizontal axis representing the degree of insulation. This allows us to visualize the relationships among the major types of policies influencing human behavior toward natural resources.

The first three policy clusters thus fall on the autonomy axis, whereas indirect incentives occupies the insulation axis. Some indirect incentives are highly insulated and have few inter-sectoral consequences. But other indirect incentives have lower insulation and can cause unanticipated impact that crosses from the focal sector to the environmental sector. If indirect incentives support the environmental sector policy, then they are not an issue. But if they have deleterious effects, then the more insulated they are, the better.

When policies do create problems across sectors, then the level of au-

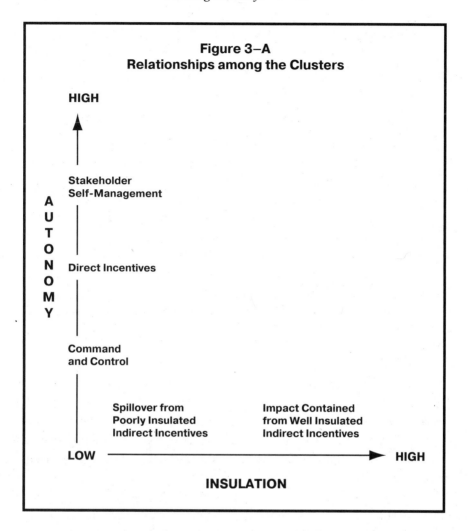

Figure 3–A
Relationships among the Clusters

tonomy of policies within the natural resources sector may be important. Highly structured policies allowing little autonomy will exhibit much greater implementation costs in settings with strong, and poorly insulated, indirect incentives pulling away from the sectoral policy objectives.

As the discussion above indicates, all four of the approaches have occurred in a wide variety of settings. Sometimes their consequences were intended, sometimes not. National policy settings usually consist of some combination of all four clusters. There may be many interactions among different policies that fit into the different analytical categories. No single policy exists in isolation—each one may solve one problem only to create another.

These four policy clusters provide a range of options and combinations for consideration. All must be examined when environmental policies are at issue. But there are also various mechanisms available for implementing a strategy based on any of the four clusters.

MECHANISMS TO IMPLEMENT POLICIES

As the discussion above implies, there are numerous avenues available for implementing public policies. Simple images of government agencies enforcing regulations do not capture the diversity of possible approaches. This section introduces thirteen different approaches to policy implementation. And even this is an oversimplification, since one mechanism—laissez faire/ market approaches—contains numerous distinct components, and others also exhibit variations on a theme.

Moreover, there may be various sequences in the use of different mechanisms. One of the options that appears early on in deliberations about policy options is the choice of an executive agency to carry out the functions.

Organizational Champion

One of the most common approaches to policy implementation is the selection of an organizational champion. A public agency, a non-governmental organization, or a newly created entity is entrusted with the task of executing the actions needed to bring a new policy to fruition. Then the chosen champion is the recipient of assistance to strengthen its capacities to perform the task.

Organizational favorites have changed through time. In the 1960s and '70s parastatal bodies were depicted as the performance favorites at the national level. They avoided bureaucratic red tape and overly political operations. Managed as enterprises, but within government, they were expected to inject resources into treasury coffers and help newly independent nations to get on their feet. But their performances varied greatly and they often became drains on the public purse while they operated outside civil service regulations and public scrutiny. Now, instead of favorite champions, they are seen as problem children and targets for privatization.

At the same time as parastatals were enjoying a honeymoon, project level organizations called Project Management Units (PMUs) were in their heyday. They too avoided the red tape and the limited administrative capacity found in many third world settings. But it turned out that they also had limited ability to generate sustainable processes, and they are now in disfavor.

Today's favorite organizational champion is the Non-Governmental Organization, or NGO. This is particularly true in the environmental field where local conservation organizations, social forestry groups, and water users associations are common. The reasons for the NGO's popularity are twofold: first, they are viewed as close to the people and therefore more likely to follow strategies that have a chance of lasting; second, they also bypass the red tape and bureaucratic inertia a la the parastatal and PMU. The jury is still out on any verdict concerning their effectiveness. But any judgement must examine their relative strengths for performing particular functions. They are *not* the answer to all prayers, and favoring them may weaken the public institutions that need strengthening for long-run success (Meyer, 1992).

Price-setting or regulation may be entrusted to a public agency, resource protection may be spearheaded by a national parks service or a conservation-oriented NGO. And mobilizing local communities to perform services may fall appropriately to voluntary organizations or political parties. But the function will suggest the range of effective organizational forms. Sometimes mixtures of the above, such as social forestry programs, are implemented by single public agencies, or a cluster of them, or even a mix of public, private, political, and ethnic organizations.

But when choosing organizational champions it must be remembered that organizations are both outcomes and causes—they are outcomes of political power struggles to determine who gets to control resources as well as what benefits will be produced, and they are causes—the choice of implementing organization partly determines who will get access to the resource (and how it will be used in the future) and it creates barriers and opportunities that program managers will encounter. The case of public sector forestry management illustrates the different biases contained in different organizational vehicles.

Public sector responsibility for forestry management and protection is commonly entrusted to one of four types of national agencies in third world countries. They are: (a) a parastatal body, (b) a department inside a larger ministry, (c) a separate ministry of forestry or natural resources, or (d) a ministry of natural resources and tourism. Each of these configurations makes a statement about the perception of the forests held by national leaders and about how access to forest resources will be controlled. (G. Honadle, 1993)

The decision to adopt the parastatal model often reflects the distribution of power in the country. When a small minority (military, racial, or tribal) holds an inordinate amount of power, a parastatal is preferred as a way of isolating control of forestry resources from popular pressures and letting a small group quietly get away with extracting the wealth. Thus, the artifact reflects the distribution of power and it also partly determines the constraints and opportunities surrounding those who manage the resource. Numerous countries in Africa and Latin America show this pattern of organizational choice and resource use.

Another common approach to forestry is to place it under a ministry of agriculture or a ministry of mines. In such a setting, agricultural production or mineral extraction gets priority attention and forestry gets lower priority in the battle for financial resources. On a positive note, the agricultural version of this configuration can promote the integration of forestry concerns and tree crops into the agricultural agenda. Multiple uses of wood products, greater distribution of the tree cover, access to extension facilities and services and various other factors distinguish this approach from that of the parastatal (in the case of agriculture). But a view of trees as a resource to be "mined" characterizes the case where minerals and mining is the dominant part of the organization.

The third major option for placement of forestry responsibility is in a separate ministry, often called "Natural Resources" or "Forestry and Natural Resources." In some situations this includes minerals and mining, which can have detrimental effects on the priority given to forestry. But in most cases, this separate setting strengthens the forestry focus.

Such ministries, however, seldom compare well with agriculture in the competition for funding, facilities, and people. But the dominance of forestry professionals within the ministry does have a positive effect on esprit de corps. Foresters are at least not second class citizens in their own organization. Their own professional norms occupy a more prominent position in the organizational mythology.

This may also lead to conflict with other organizations. For instance, protecting local hardwood forests from depletion may require confrontation with foreign exchange earning extractive industries (such as mining) or wood energy intensive agricultural activities (such as tobacco production). Or it may pit foresters against national elites bent on transforming local natural wealth into foreign financial deposits.

This organizational form allows conservation perspectives to attain an equal status with production perspectives. But poor links to village organizations and an underdeveloped extension system often characterize this model (in one country a local forester called the forestry extension system a "tree without roots").

The fourth major location for the public sector unit charged with responsibility for forestry management is in a ministry of "Natural Resources and Tourism." This tends to link trees and wealth in a very different way than the other options. When countries have unique, special, or abundant fauna that attract global attention there is often a tourist industry based on that resource. Tourism generates foreign exchange through the preservation of the resource rather than its extraction (although the ivory traffic and the trade in endangered species do represent illicit, short-term extractive behavior). Thus there is at least some pressure toward afforestation, species conservation and treatment of indigenous species of flora and fauna as national resources.

Although the use of an NGO rather than a government unit is often cast as a radical departure from previous practices, it is actually no more than an option within the organization champion strategy. It simply broadens the choice. The organizational champion approach just chooses the most appropriate implementer and helps it to do what it already does. Sometimes that may be an NGO.

The organizational champion approach also may involve designing a new agency to implement new policy. The commotion surrounding the structure and placement of organizations charged with the execution of National Environmental Action Plans (NEAPs) attests to the importance attached to the choice of a champion and to the frequency that a new agency is designed. Thus the champion decision may be the first in a sequence of actions leading to policy implementation and to the application of other instruments.

Bureaucratic Reorientation

The organizational champion approach assumes the availability of choice. Bureaucratic reorientation assumes that the organizational mechanism is a given, that a public agency is the predetermined implementation vehicle, that it is well below optimal in terms of the demands of a new mission, and that major initiatives will be needed to transform it into a properly performing entity. Redesign of procedures and structures, retraining of staff, enhancement of facilities and equipment, and technical assistance are all integral parts of bureaucratic reorientation. But informal organizational missions can complicate the reorientation process.

The formal allocation of power and authority does not always reflect true decision making practices. The Bureau of Forest Development (BFD) in the Philippines shows the importance of informal factors. Although its formal mission, or espoused policy, was to protect the forests and regulate logging, its informal operations, or policy in use, emphasized assessing royalties on illegal loggers and protecting logging interests. Under Marcos its staff lived in comfort based on the bribes they extracted. In attempting to clean up this situation the Aquino administration encountered the need to dismantle the BFD clique. Senior people were retired, some junior people were dismissed, and most of those who remained in the civil service were transferred to other organizations. In addition, a new department became the home of the previous BFD functions. Thus both formal reorganization and informal personnel clique-breaking strategies were used to reorient a predatory bureaucratic entity and break some of its ties to external influence (see Dauvergne, 1997). In this case the task was to close the gap between the espoused policies and policies in use. In other cases, reorientation will involve substituting new espoused policies for old ones and then setting out to minimize the gap between the formal and the informal.

There is also a counterintuitive element to bureaucratic reorientation. It is more likely to work in an organization that has high capabilities but marginal affinity to the mission than in an organization that has an ideological predisposition toward the mission but low capacities. That is because capacity is redirected, not created.

This was identified in a study of US organizational responses to the 1973 Environmental Protection Act requirement for environmental impact statements (Clarke and McCool, 1985). The high performers in this case included organizations not known as environmentally-conscious. Likewise, in the Philippines, work that led to the coining of the term "bureaucratic reorientation" (Korten and Uphoff, 1981) occurred within the National Irrigation Administration—a very high-capacity organization (see Korten and Siy, 1988) that was not predisposed to support local participation in its mission. And yet it became one of the success stories of development administration.

Although the use of the term "bureaucratic reorientation" has diminished over the last decade, this approach is likely to experience a revitalization. The reason is simple—as ecological crises increase, the need for rapid response also increases creating pressure to use existing organizations rather than inventing new ones. In fact, some recent publications are inadvertently highlighting reorientation capacity. For example, the United Nations Development Programme's *Human Development Report* offers indicators such as the ratio of teachers to soldiers in a country. The intent is to show relative priorities and non-productive resource use. But it also helps to identify excess capacity organizations that could be mobilized temporarily to perform more positive tasks. These may be candidates for bureaucratic reorientation in the pursuit of environmental policy objectives.

Environmental Dispute Resolution

There are times when lawmakers and policy makers are unable to decide what must be done. Just as standoffs can occur in labor disputes, so too gridlock can characterize environmental disputes. Sometimes resolution occurs in courtrooms, but often the parties hesitate to leave the decision to judges—they fear the result or the process. These situations have given rise to an approach that goes under the labels of "environmental mediation," "environmental dispute resolution," (EDR) or "alternative dispute resolution," (ADR). Similar to arbitration, it uses outsiders to bring warring parties together. The experience base is long and growing (Amy, 1987; Bingham, 1986; Crowfoot and Wondolleck, 1990), it has been used for both policy formulation and site-specific decision-making in developed nations, and it is increasingly offered as an option available to international donors, third world governments, third world communities, and regional institutions.

The characteristic signature of these processes includes the following:

- voluntary participation by the involved parties;

- direct interaction among the parties in group or "workshop" type settings;

- willingness to abide by mutual agreements or consensus decisions;

- agreement on an acceptable process; and

- use of an outside facilitator to guide the process.

This collection of approaches has also been advocated as a way of settling disputes among sovereign nations. And many recognize its potential application in situations where sovereign states confront international institutions on unequal playing fields.

The market for ADR skills is growing, but as is often the case with rapidly expanding markets, service suppliers vary greatly in their abilities. There are reputable institutions, such as the Keystone Center, that have been involved in these exercises for years, and there are development specialists who have developed and honed third world tested conflict resolution methods over the decades. But there are others, often unemployed organizational development experts, who now see the market and claim ADR expertise.

One of the positive side effects of ADR exercises is the building of organizational capacity within previously weak organizations, such as citizens groups. Studies of these exercises indicate this is an important outcome, and given the present emphasis on non-governmental organizations in the environmental sphere, it might have great potential in the international conservation movement.

Markets

Some writers contend that just leaving everything alone and allowing the shortage of environmental "goods" to accelerate will produce price rises that will adjust for the value of environmental services. But few policy analysts seriously hold such faith in the timeliness of market response. Most advocates of market-based policies propose government intervention to hasten market adjustment and even create new markets. The key characteristic of the market strategy is that supply and demand, rather than administrative fiat, is used to adjust the production, consumption, and distribution of goods and services.

Market-based implementation mechanisms employ charges, rights, information, and rewards to influence the supply and demand for environmental goods, bads and behaviors (Anderson and Leal, 1991; OECD, 1991). There are eight readily identifiable market-oriented approaches to the implementation of environmental policy.

The first of these is the erection of a schedule of *pollution charges*. Such charges penalize actors whose activities result in emissions that degrade air, water, or soil resources and they therefore raise the prices of the goods produced by the polluters. Pollution charges often accompany direct regulation and they promote technical innovation to reduce pollution levels and keep producers competitive. There is a major administrative dimension to pollution charges, however. Monitoring capacity is a necessity. And emission charges are most easily enforced when applied to pollutants emitted in large quantities from large stationary sources.

The second market approach involves *product charges or credits*. Charges can be applied to products that release polluting substances during consumption (such as solvents), or they can be used as a tool in energy pricing. And product charges can be used as a proxy for emission charges when pollution is diffuse, that is when there are numerous small or mobile sources.

Product credits, rather than penalizing polluters, reward non-polluters. And credits can create markets that were previously stifled. For example, in Minnesota, before the monopolistic utility was required to credit small producers of electricity, there was little market for household wind power. Either the household had to produce all of its power or it was simply contributing to the grid without compensation and it could not even recover the investment in equipment. But the law requiring credit to co-producers created a market for household wind power.

The third market mechanism is *a deposit-refund system.* This is useful for either generating a market for recycling materials or for ensuring efficient collection of goods that can release pollutants in the event of inappropriate disposal. Examples of the latter include oil products, herbicide and insecticide containers, refrigerators, batteries, air conditioners, and fire extinguishers. Refund systems have even been used in urban areas of Brazil as a substitute for public garbage collection.

A fourth mechanism is a *tradable permit system.* This is an alternative to pollution regulation that is less administratively costly than regulation and it has the added advantage that it promotes innovation. By limiting the levels of a pollutant, but not telling *how* to do it, producers are rewarded for adopting innovative technologies. In Singapore, for example, the amount of automobiles allowed downtown during specified hours is limited by a permit quota. But such permits may be sold, thus creating a market for them while constraining traffic and air pollution.

Permits are also a common feature of water markets. Members of irrigation systems and watersheds can sell water rights to the highest bidder or keep it for their own use. This promotes efficient water use and such markets have been in existence for many decades in Europe, Asia, and North America (Maas and Anderson, 1976).

Although it is a more recent innovation, and one largely confined to wealthier nations, tradeable development rights can play important roles in land markets. In the United States, some states, such as Pennsylvania, Maryland, and New Jersey allow the purchase of development rights to agricultural land. The purchaser of the rights can then exceed density allowances (by a specified amount) when building on land with development rights attached. The owner of the agricultural land then receives a large cash payment as the sale value (and tax value) of the land plummets. But the owner keeps the land and the integrity of the landscape is preserved. And as the value of development rights rises, purchasers of such rights can sell them on the open market rather than using them themselves. Alternatively, public or private purchasers can retire the rights without using or trading them at all.

With the rapid increase in urbanization worldwide, and the trend toward increasing freehold land tenure, the potential for policies establishing tradeable development rights is growing. Indeed, tradeable permit systems are also politically palatable in settings where those in power are threatened by more radical environmental policies.

Fifth on the inventory of market-based approaches is *consumer informa-tion provision*. This is often called "green labeling." It basically is allowing consumers to make more informed choices about the products they buy by identifying the environmental responsibility exercised by the producer.

Labels may identify constituent ingredients, production processes or sources of raw materials (e.g., not from mature tropical rainforests). But some form of certification system, impartial monitoring, and reporting is needed for this to work (Upton and Bass, 1996). Otherwise it quickly degenerates into little more than competitive obfuscation.

The sixth mechanism is *market access control*. In the case of the purchase of goods and services by public bodies this amounts to the creation of rules for competition for procurement contracts. Providers of consultant services or commodities to bilateral agencies, regional development banks, the World Bank or host government units through donor contracts could be required to demonstrate their environmental responsibility.

Such markets have previously been constrained in terms of geographic source and type/ownership of bidding organization. There is no reason that new criteria related to sustainable resource use could not be developed and applied in such a setting. It could promote innovation in the same way that other mechanisms noted above do. And the total market created would ap-proach twenty billion dollars. If domestic procurement by government agen-cies worldwide also were characterized by such an approach, the market size would soar far beyond this level.

The seventh approach is *alternative income competition*. This is basically a scramble to develop a market for benign use of a resource when there is already a market for destructive use. For example, ecotourism and extractive reserves are attempts to generate benefits in places where deforestation, agri-cultural or urban encroachment, mining, or other harmful activities are either planned or already underway.

But this is not as easy as it seems. Alternative income is often both lower income and income for less powerful people. And market linkages can be difficult and costly to establish and they also may be unreliable. Nevertheless, this is a common endeavor at the turn of the century, undertaken both by NGOs and by international donors.

The final market-based mechanism is *clear property rights*. The rationale for this approach is simple—rights can be traded, producing a market. And it is desirable to know exactly what you are purchasing. This is actually an old approach. Land purchase organizations such as national trusts or the Nature

Conservancy have been practicing this for a long time. But it also applies to intellectual property rights. And in the form of "biodiversity prospecting" (Reid and others, 1993) it is extending to rights over the medicinal or industrial applications of natural substances.

Unfortunately, the clear property rights mandate has achieved the status of panacea in the eyes of some observers (such as Panayotou, 1993). Third world systems of land rights are especially befuddling to western observers. The equivalent of freehold land tenure is advocated by many. But in some circumstances (the Amazonian frontier in Bolivia, for example) clear property rights have led to the creation of a pioneer class that moves in, cuts down, sells, and moves on, leaving degraded land with secure rights in the hands of exploiters.[2] Clearly, there is more to it than specificity of rights. The nature of the rights is important, too, and we may be moving into an era when freehold rights comprise more of the problem than the solution. This is also the case in the United States, where the "wise use," or "takings," movement has latched on to property rights as the key to blocking the implementation of environmental policy. Takings advocates promote the idea that government should compensate landholders for any loss of sale value or economic use resulting from the enforcement of environmental policies, and the concept and tradition of freehold land tenure makes their arguments seem more reasonable than they may be.[3]

Freehold land tenure bundles most of the rights to real estate and allocates them to a single actor—the freeholder. For example, the rights to live on a property, cultivate it, plant trees on it, harvest trees from it, hunt game on it, travel across it, extract minerals from it, use water that passes over it, invest in it, sell or otherwise alienate it, inherit it, build on it, or allow others to do any of these things are all separate rights that may be vested in different people or groups or institutions. But under the freehold system they are mostly bundled together and vested in the owner. Mineral rights may be separate; zoning may constrain some activities; the water may be state-owned while the land beneath it is private; and the right to block people from crossing over it may be lost over time, but generally, the freeholder has the predominant control over the land. This leads to the perception that land is either public or private with no combinations of intermediate ownership types.

But nothing could be further from the truth. In many parts of the planet, many actors share these different rights. And who has what rights varies considerably from place to place. Each twig can be taken out of the bundle and considered separately. And such consideration leads to different resolutions of land use conflicts than those promoted by takings advocates. This importance of differences among local tenure systems will be examined more fully in Chapter Four in the section on resource decision systems.

Policy Pronouncement

There are times when simple pronouncements of new policies make a difference. For example, when the President of Malawi announced a higher price to be paid for rice in one district of the country, it stimulated rice planting in that area. Similarly, changes in hiring practices, procurement rules, and other initiatives can be introduced at the stroke of a pen when the element to be changed is under the direct control of an administrator or official.

This does not mean that no preparatory or follow-up work is needed. New procurement policies, for example, will need new bidding documents and possibly a new review process to work. Nevertheless, a pronounced policy change can set the wheels in motion without recourse to lengthy and cumbersome legislative deliberations and procedures. Issues of public land management, prices and fees, and internal bureaucratic operations lend themselves readily to this approach. They can be very important for establishing the rules that guide the competition among private or NGO actors and they can affect the impact of government operations on the environment.

International treaties and accords also have the effect of policy pronouncements. Once they are signed by the representatives of a nation they are espoused policy. Signing may be preceded by legislative approval or it may be the sole prerogative of the executive branch. But once the required number of signatories have endorsed an international agreement, it becomes internally binding. Such policies often focus on transborder problems or the depletion of world commons resources.

Another, promising, type of policy pronouncement is found in the Netherlands. Here private firms that voluntarily participate in environmental and sustainable development programs receive some latitude in meeting certain pollution targets or other regulatory objectives. This allows experimentation, innovation, and progress toward public goals. Agency heads often have the discretion to initiate such programs. In fact, most policy pronouncements are based on executive discretion that needs no legislative approval.

Legislation/Decree + Regulation

Regulation is the mechanism most often associated with environmental policy in many people's minds. It has much in common with the organizational champion approach since a regulatory agency can be the unit responsible for overseeing and controlling the activities of others.

But regulation is also different from the champion approach in that regulatory bodies do not generally *do* things—they merely guard against others doing things. They are watchdogs, not sled dogs. They have a command and control function, not a production function. Indeed, they are the archetype of behavior blocking.

By itself, regulation is generally a costly approach to environmental policy. It is costly both in money terms and in political capital terms. Regulators do not make friends and regulatory agencies are seldom popular. But regulation does fit well into very hierarchical, authoritarian cultures. In such settings, regulatory agencies are just the arm of the "big man."

Regulation of some sort may also be a necessary complement to other approaches. For example, water markets or tradeable permit systems may need watchdog agencies to work. And enforcement of maximum pollution levels or game laws may be crucial components in broad-based environmental initiatives. Likewise, regulation often accompanies monopoly. When a private organization, such as a utility, is given a monopoly over the provision of a service, or a mining company is given sole access to a resource, regulation goes along with it. Independence is sacrificed for monopoly power.

The creation of regulatory functions is also the creation of opportunities for corruption. And this has been especially true in former colonial territories. Foresters in some countries, for example, make most of their income from bribes. This supplemental income dimension is exacerbated in poor nations where public salaries are too low to ensure family survival and where cultural norms support tribute for services. But there is another factor that also defines the nature of regulatory agencies in third world settings—weak capacity.

The indicators of this weakness are multiple. They include: no access to laboratory facilities, inadequate budgets, poor training of staff, lack of organizational allies when confronting adversaries, and extremely high percentages of budgets going to the wage bill. In some cases, Papua New Guinea's Department of Environment and Conservation, for example, the regulatory function is combined with a public land management function, thus diluting the focus and the application of already scarce resources.

Given all its associated problems, regulation still remains a major tool for environmental policy implementation. It is just an unruly tool for a difficult job.

Devolution of Ownership/Management

An alternative to central control or direction is the devolution of authority or ownership. This has a long history. Sometimes it was accomplished in a de facto and informal way; other times it was a formal devolution. An example of the informal is the coexistence of multiple systems of traditional land tenure within a country. The formal designation of "traditional lands," or "tribal trust," or some other category indicates great variety from place to place without central control. In Africa, many nations also have a parallel court system, with traditional courts overseeing village life and applying traditional rules that differ from place to place. At the same time, the "modern" courts enforce criminal laws and guide the national economy using commercial codes that apply nation-wide.

The devolution of control of a natural resource base to a local community or group is also common. Community resource management programs in Asia often focus on forests or fisheries. In Africa, a combination of forest and wildlife resources are increasingly controlled by local areas. The CAMPFIRE program in Zimbabwe is one of the most famous examples, but there are many more.

Although this is an increasingly popular mechanism because it has the advantages of administrative efficiency and the promotion of local self-reliance, it may require complementary actions to succeed. Without national policy settings that support markets but protect localities, and/or local community capacity-building, devolution can turn into disaster. And, just as the privatization movement has often resulted in the privatization of non-viable enterprises creating what has been called "lemon capitalism," so too, natural resource devolution faces the danger that the resources devolved will be marginal ones.

Given the caveats, however, this trend is likely to continue and perhaps even increase. One trend that supports it is the desire to conserve cultural diversity. Many cultures that are peripheral to the world economy need resource reserves to survive. Since natural resource management is central to many of these cultures, devolved control is a tactic for improving survival chances.

Another factor favoring increased devolution is the limited capacities of many national governments. They simply cannot treat their territories as command areas because they do not have the wherewithal to manage them. And the worldwide shift toward market economies also supports a devolutionary approach to natural resource management.

Public Trust

There are two basic approaches to the creation of public trusts. They are financial trusts and physical trusts. Both are noted below.

Financial trusts require seed capital to begin the endowment as well as an institution responsible for managing both the principle and the income it generates. Often linked to the resources of international NGOs (such as in Bhutan) or a debt-for-nature swap to accumulate the initial capital, the financial trust uses a foundation or grant-making entity to manage the use of the income, or, in some cases, to spread out the expenditure of the principal over an extended period, such as twenty years.

Trusts can take many forms—revolving funds, endowments, and extended-time income sources. Funding can be entirely up front, or revenue sources such as a "conservation fee" assessed on tourists in Belize, or a possible environmental levy on mining in Papua New Guinea, can augment initial finances. What all these variations have in common is that they establish a stable source of long-term funding for efforts to conserve biological diversity and promote sustainable natural resource use.

According to Barry Spergel of the World Wildlife Fund in Washington, DC:

Trust funds and endowments have recently come to be used as a biodiversity conservation tool. World Wildlife Fund (WWF) has assisted (or is currently assisting) in the establishment of national level conservation trust funds in Bhutan, the Philippines, Papua New Guinea, Belize, Guatemala, Honduras, Colombia, Poland, Uganda, Nepal, Mexico, the republic of the Congo, and Namibia, as well as two multi-national trust funds—a tri-national fund in Poland, Slovakia and Ukraine, and a binational one for Poland and Belarus. Of the fifteen conservation trust funds that WWF has been involved with, six have been associated with Global Environmental Facility (GEF) projects; three have been done in collaboration with the US Agency for International Development (AID); and one is now being proposed for funding by the Asian Development Bank. The other five WWF-assisted trust funds have not been associated with any particular outside donor agency, but have been developed entirely as in-country local initiatives, which have then sought outside funding. (Spergel, 1993:1)

A *physical trust* involves the establishment of a renewable energy source to take the place of a non-renewable energy source that is in the process of depletion. For example, a coal mine could set aside a percentage of its profits so that, upon the exhaustion of the coal vein, a mature forest would have been created. The forest size would be determined by the annual level of British Thermal Units (BTUs) contained in the ore yield—the forest would need to be large enough so that it could produce a sustainable yield equal to the annual energy production of the mine.

This concept, originated by Herman Daly (Griesinger Films, 1991) could be applied to project investments of international donors. Or it could be used by national governments to set conditions for international corporations extracting exhaustible resources. In either case, space would need to be reserved so that the physical trust could be managed for the public good. The physical trust does not necessarily require a new institution to work. In the example noted above, an already existing public land management agency or forest department could be given implementation responsibility.

Debt-for-nature Swap

Debt-for-nature swaps often go hand-in-hand with the establishment of public trusts. This is simply a funding mechanism to create the financial endowment and to set up a conservation area that the endowment will support. It grew out of the huge debt burden accrued by third world nations in the 1970s and 1980s (Miller, 1991).

The idea originated with Thomas Lovejoy of the Smithsonian Institution and has been carried out by a number of NGOs, most notably Conservation International and the World Wildlife Fund. The underlying idea is simple—if commercial banks in the industrial nations see their lending portfolios to third world nations as at risk, why not buy up the debt in dollars at a discount, get local debtors to pay in local currency and/or public land, establish a local nature preserve with an endowment, and thus satisfy the needs of numerous actors?

The international NGO then gets agreement from both the bank and the third world country, raises the funds to purchase the debt in hard currency, works with the local government and local NGOs to establish the endowment and preserve, and then either it pulls out or maintains a role in the conservation activity. International donors often play supporting roles in these swaps, but sometimes they are entirely out of the picture. The role of match maker, however, is crucial and is typically played by an organization that operates internationally.

Unlike the debt-for-nature swap, which has its origins in the creativity of international thinkers, some mechanisms have been mainly rooted in internal institutions. One of these is an exhortatory approach to policy implementation that relies on political persuasion to obtain results. The source of this approach was mainly religious pulpits and independence struggles.

Political Exhortation and Mobilization

The idea that speeches alone can reorient economies seems naive today. But in the 1960s, fresh from the political contests that resulted in national independence, many leaders of new nations relied on charisma and political rallies as policy implementation mechanisms. They served them well during the independence struggle and they carried over into the nation-building effort. And, indeed, no one who heard the speeches of the young Julius Nyerere, or Soekharno, or Michael Manley, or Kwame Nkrumah, or Ferdinand Marcos, or Mao Zedong, or Jomo Kenyatta could doubt the ability of a fiery leader to prod people in new directions. The political pulpit of the nationalists, like the religious pulpit of the colonialists and missionaries before them, served as a policy tool.

The tendency to impose single-party regimes reinforced this tactic. Since the one party was the only legitimate political voice, and since the leader of the country was also the head of the party, party organization became a major policy implementation mechanism. Mobilizing the countryside for economic development would occur through the application of two parallel organizational devices—the national government agencies and the political party apparatus. The agencies would implement the national programs. The party would be a watchdog over the agencies and it would mobilize popular action in support of agency programs. At least, that was the plan.

But the post-independence experience proved to be more complicated than expected. Exhortation was often at odds with the signals sent by the marketing system. When the words of politicians promoted harder farmwork and planting more area in certain crops or replacing trees, but the message of low crop or fuel wood prices contradicted those words, the price signal dominated farmer response. Declining production in the seventies and eighties was linked to the stagnation of farm gate prices.

This lesson should not be lost on those bent on the implementation of environmental policy. Indeed, much discussion is based on the premise that if people only knew the impact of their actions, they would change them. This leads to "conservation education" and exhortation as a strategy for influencing and implementing policy. But, as we have already seen in this study, exhortation alone is seldom a sufficient incentive. The web of direct and indirect incentives surrounding behavior is crucial and cannot be ignored. Even persuasion plus a weak incentive may not do the trick.

An illustration of this comes from the United States. The major limited access arteries leading into and out of major cities often contain traffic lanes reserved for high occupancy vehicles, or HOVs. The idea is to reward and promote ride-sharing during the daily commute to work and back home to the suburbs. But the HOV lanes remain underutilized and they actually contribute to the congestion on the highways. New Jersey has come to the conclusion they should be abandoned, and some Minnesotans see them in a similar light.

No matter how many TV ads, radio spots, billboard messages, or other exhortations greeted commuters, most rode alone. The reason was simple— the time and convenience saved by using the HOV lane in the middle of the commute were less than the amounts lost at the two ends by car pooling. The incentive that accompanied the coaxing was outweighed by disincentives imposed by the existing settlement pattern. The management of a car pool had its own costs, and exhortation combined with weak incentives could not overcome this fact.

There is another approach similar to persuasion, but it is more data-based and informative than it is exhortative. This is publicity and public awareness.

Publicity and Public Awareness

Most approaches to policy implementation focus directly on the management of human behavior. But the approach we are calling "publicity and public awareness" does not. Instead, it attempts to create the conditions that will foster new behavior. It focuses on an enabling environment for policy implementation. There are five basic variations on this theme.

The first variation is exemplified by the strategies of international conservation organizations such as the World Wildlife Fund (WWF). WWF does not engage in resource ownership and management (as done by the Nature Conservancy), or use litigation (as is done by the Natural Resources Defense Council and Environmental Defense Fund) or engage in direct confrontation (as practiced by Greenpeace). Instead, the approach is educational. By raising public awareness of the benefits of wildlife and strengthening an appreciation for other species and natural habitats, WWF expects to influence public opinion and eventually affect policy. Direct interaction with policy makers also follows an educational tack rather than a confrontational one.

This approach is based on the assumption that knowledge leads to attitude change, which, in turn, leads to behavior change. Publicity campaigns, supplying materials to the educational system and incorporating ecological questions into school certificate exams, use of mass media, and studies and publications are all elements in this approach. It is, in some ways, the environmental version of the strategy underlying agricultural extension. Indeed, government ministries employing a public awareness strategy often borrow insights and techniques directly from the agricultural extension system.

The second variation on the publicity and public awareness theme is information intensive in a different way—it assumes that the generation of data about the environmental consequences of specific development activities will induce confrontation between different interest groups and lead to revision of harmful policy applications. This is the Environmental Impact Statement, or EIS, approach.

Requiring an EIS before dams are built across rivers or roads are cut through forests generates public access to data about the impact of these projects on the environment. It allows deliberation in a public forum before the action is taken. Such public deliberation is expected to lead to modifications in poor projects, cancellation of bad ones and support for good ones. Sometimes it does. But this is not always the case. The administrative requirement for an Environmental Impact Statement does not always generate data that are used in public debate, especially in informationally closed societies or organizations. And when the power balance among pro and con organizations is lopsided, the ensuing discussion scarcely resembles debate.

This is, however, a major approach to policy implementation that is found in third world countries, in the United States, and among the international donor organizations that sponsor third world development. Indeed, it is also a primary theme in the environmental literature.

The third public awareness approach combines elements of the two above. This approach generates ecological data focusing on a geographic area. Studies using Geographic Information Systems (GIS) and inventories of biological diversity are examples of this. The assumption is that the availability of information about fragile ecosystems, endangered species or endemic resources will discourage the beginning of destructive activity because potential developers and miners will see that the mobilization of public opposition, the call for an EIS, scrutiny of bureaucratic decisions and decision-making processes, and bureaucratic foot dragging are all facilitated by this information. Indeed, this approach is not limited to third world settings. The County Biological Survey in Minnesota has already had such an effect.

The fourth approach is represented by a public planning, negotiation and consciousness-raising process. National Environmental Action Plans (NEAPs) need to be mentioned here. Although NEAPs are not really separate implementation approaches, they are information-generating activities that mold public and bureaucratic opinion and thus attempt to create settings conducive to effective environmental policy implementation.

Likewise, public fora help to aggregate, mobilize, articulate, and alter public opinion. The UNCED conference(s) in Brazil in 1992 drew international attention to environmental and sustainable development issues and built public support for policy change. A plethora of follow-up congresses and task forces have followed the Rio summit. They are inducing and easing the way for policy change.

In the United States, the President's Council on Sustainable Development represents this public consciousness-raising strategy. Many community-level efforts and state-level deliberative efforts are underway. In Minnesota, a one-week Sustainable Development Forum sponsored by the Attorney General's Office and an Environmental Quality Board-sponsored two-year-long Sustainable Development Initiative brought together business, environmental, academic, government and Native American representatives to focus on the sustainability of present trends in such areas as forestry, manufacturing, energy, recreation, agriculture, land use and settlement, and mining. This was then followed by a Governor-appointed, state-level, thirty-member, Sustainable Development Round Table to refine policy recommendations and to establish action agendas (Minnesota Roundtable on Sustainable Development, 1998).

Educational initiatives that incorporate knowledge about environmental history, sustainable development principles, human population growth and its effect on resource scarcity, and conservation biology into the public primary and secondary school classrooms all fall under the awareness approach, too. Even cable and public television programming fits here. And voluntary organizations also sponsor such initiatives. The Izaak Walton League's "carrying capacity" and "sustainability" efforts are examples of this.

A fifth approach within this strategy is also found in wealthier countries. It uses the private lecture circuit. Corporations and public institutions offer a market for entrepreneurs who "sell" prescriptions for the future. Numerous institutes of "sustainable this-or-that" exist all across the American landcape, for example. Chambers of Commerce, professional organizations, community fora, national, state, and local government institutions, private corporations, non-profit organizations, and foundations all present a platform for issue advocacy.

This can also generate training programs and spawn local affiliate organizations. Some may be little more than wishful thinking and alchemy, but others indicate a potential renaissance in public thinking. One of the newer ones, for example, is called the "Natural Step." Begun by a Swedish doctor, this perspective is based on what is characterized as what Swedish scientists can agree upon as objective descriptions of human impact on nature. They are phrased as commandments. The first three are derived directly from scientific consensus. The fourth combines a normative statement about preferred social structure with a statement about the need to improve energy efficiency.[4]

Publicity and public awareness initiatives, then, come from both public and private sectors. The role of public policy is to facilitate both types.

Subsidy

Subsidies are often used to induce investment in energy efficient or environmentally benign technologies. For example, subsidizing the cost of home insulation, wind generators, or solar collectors can lower demand for electricity and have desirable environmental impact. And, indeed, these are among the more effective types of subsidies.

One problem with applying credits or subsidies is deciding what to subsidize. Experience suggests (see OECD, 1991) that it is best to use this mechanism on "front-end" technologies rather than "end-of-pipe" approaches. The former support the shift to less harmful technologies, whereas the latter can contribute to a longer use of the more damaging ones.

Another problem, well documented in the development literature, is that subsidies can work to thwart the objective of sustainability.[5] When agricultural inputs were subsidized and then the support was removed, use declined. And keeping the subsidy generated a recurrent cost problem. Thus subsidies can create artificial demand that either culminates in abandonment when the subsidy is removed, or results in financial burdens when the source of funding declines.

Environmental subsidies will most likely encounter the same problems. Indeed, due to this experience plus a general reluctance to recommend subsidies on the part of neo-classical economists (because of the price distortions and economic inefficiencies that they can produce) subsidies are not a favored economic instrument for environmental policy. And international trade agreements frown upon them.

The Organization for Economic Cooperation and Development (OECD), however, does distinguish some situations where subsidies may be useful:

As a general rule they are incompatible with the Polluter Pays Principle. . . . However, subsidies remain a form of economic instrument which can be effective in certain circumstances, such as payments for positive externalities, cleaning up of derelict sites, catching up pollution backlog. It is also generally accepted that the revenue of pollution charges may be earmarked to achieve environmental goals. (OECD, 1991: 15)

In poor countries, subsidies are most useful for inducing household-level investments in preferred technologies. This is the social level where financial fragility can limit alternatives most severely. In more industrial settings, the firm may be the target of opportunity. But, as a component in a strategy to convert society from one set of technologies to another, subsidies may have a major, if temporary, role to play at all levels of economic development.

Taxation

Taxation as an instrument for the implementation of environmental policy is both a derivation and a departure from neo-classical economics. From the neo-classical perspective, there are four considerations that are key to evaluating the impact and desirability of a tax. They are:

a. *Revenue* (how much money will be raised by each alternative?);

b. *Administrative ease* (how difficult and costly will it be to collect the tax?);

c. *Equity* (who bears how much of the burden and how will it be shared among income groups [vertically] or between ethnic/gender/occupational/geographic or other categories [horizontally] of the population?); and

d. *Preference neutrality* (does the tax introduce a distortion by altering economic decisions or does it focus on a good or activity for which demand is relatively inelastic?).

The emerging sustainability focus accepts *a* and *b* above, but there is a divergence concerning *c* and *d*. As for equity, the calculation will be expanded to include impact on different natural resources, not just humans. Indeed, the distribution of the impact on the natural setting will be so important it could be depicted as a fifth focus.

As regards preference neutrality, there are two aspects of the different view. First, there is a bias toward using a tax to internalize externalities and bring the price of a good or practice more in line with its true cost. Thus, full-cost accounting needs override preference neutrality. This can be incorporated into neo-classical economics.

But the second reason for rejecting preference neutrality does not fit within the neo-classical perspective. That is a behavioral view extracted from other social sciences, such as psychology. From this perspective, there is nothing sacrosanct about not introducing a "distortion." There is no reason to believe that a market-based stimulus in the absence of a tax is any more natural than one incorporating a tax. People respond to stimuli, period.

From this second view, taxes are legitimate tools to use to induce new behavior by altering the prices attached to alternative actions. Preference neutrality is not a consideration. Behavioral impact is *the* consideration.

The relevance of this view is greater today than it was thirty years ago. During the sixties and seventies a major consideration in the tax systems of third world countries was administrative ease. Although per capita or "hut" taxes were left over from colonial regimes, the major revenue source was import/export duties. Taxing goods at the border was less dependent on literacy and required less administrative infrastructure than an income tax, corporate tax, sales/value-added tax, or user fee. And controlled marketing systems allowed the government to extract income from rural peasants without developing local-level tax management capacity—price controls combined with credit and input provision through official channels made local-level tax management unnecessary.

But today, newly industrializing countries and liberalizing economies present much higher levels of human capital and much more complex economies to the tax collector. For example, increased freehold land tenure has made real estate taxes possible. Indeed, donors have assisted tax reform and local revenue generation projects from the Philippines to Malawi. And the rise of the multinational corporation has made new approaches possible. The role of tax instruments for environmental policy implementation is growing. The "carbon tax" is just one of many formulations of this approach. Indeed, the concept of "green fees" has taken off and awareness of the potential use of taxes to guide economies in more sustainable directions is growing at an exponential rate (Von Weizsacker, Lovins and Lovins, 1997).

In the United States, 462 provisions for ecological tax penalties, credits, or other incentives have been documented in the fifty states. Divided evenly between incentives and penalties, environmental taxes are being used to achieve a wide range of policy objectives including promoting new technologies, providing insurance against environmental risk, investing in natural capital, reinforcing new markets, fostering institutional change, promoting consideration of long-term sustainability in waste production and management, and public education (Hoerner, 1998). In the future, taxes will become major instruments for changing the relative costs of different behaviors.

Trade Restriction

A basic tenet of neoclassical economics is the value of free trade. The elimination of barriers at national boundaries is supposed to bring all economic actors into the marketplace, smooth out distortions in locally protected markets, and benefit everyone by allowing comparative advantage to rule. But this is under assault today. There is a new awareness that free trade may work against sustainable resource use.

When actors in one country use production technologies that destroy habitat, the ozone layer, air quality or deplete a resource such as an aquifer, they may be able to do it for less short-term financial cost than other actors, in other countries, using environmentally benign technologies. With a free trade regime, the lower cost product can out compete the more sustainable product in the short run. Thus, free trade can accelerate the mining of resources and the subversion of strategies for sustainable development. This prompts a new look at the impact of, and roles for, tariffs, quotas and restrictions on capital flows.

Tariffs are taxes imposed on goods that cross borders. Generally they are placed on goods coming into a country, but they are also sometimes levied on goods leaving a jurisdiction.[6] Traditionally, import tariffs were often used to tax luxuries or to raise the price of goods that would compete on the local market for goods that were locally produced. In developing countries tariffs were also justified by an "infant industry" argument. That is, protection was given to new industries that had not yet climbed the learning curve to the point where they were efficient enough to compete with more mature, external producers.

From an environmental perspective, however, tariffs can be used to penalize goods that are produced in unsustainable ways or originate from threatened sources. This protects locally produced goods that follow sustainable production practices but cost more.

Quotas are limitations on the amounts of certain goods allowed to cross a border, often from a particular source nation. The limitations could be linked to factors other than just source, however. For example, if monitoring data showed that the forest cover of a particular place was increasing, then the allowed amount of timber could increase. Or, allowable levels of a product might be linked to public investment in protected natural areas and the enforcement of the protection. This would, of course, need adequate monitoring as the problem of laundering goods by routing them through third countries would exist.

Capital limits are restrictions on the amount of money (either hard currency or electronic credits) transferrable from one place to another by an actor in a given time period. In the international capital markets and financial conduits that exist today, supporting resource-depleting activity for short-term gain is easy. When resource protection policies are scheduled to go into effect on a certain date it creates a window of opportunity for exploitation. The ease of capital transfers makes it easier to enter that window. And the ease of moving capital also contributes to the ability to avoid national laws by moving money offshore or through a string of countries to bypass restrictions.

MATCHING MEANS AND ENDS

The implementation mechanisms noted above may appear as discrete interventions aimed at single behavioral objectives. Or, they may appear as components of broad-based policies aimed at promoting systemic change. The means introduced in this chapter need to be matched with the ends presented in Chapter Two. Figure 3–B displays uses for the different mechanisms. It identifies which mechanisms might be useful for addressing different desired consequences as part of a strategy using either of the four policy clusters. What this suggests, however, is that various combinations of mechanisms may be created to design a program aiming at the different behavioral objectives. Matching means and ends may involve either simple or complex approaches.

Simple approaches may make sense when a discrete behavior by a limited set of actors constitutes the threat to a site-bound resource. But if the threat is mobile or far removed from the item to be protected, then the solution will also mirror that fact. For example, international capital invested in a fishing fleet that is exhausting an endangered specie may not be susceptible to a simple strategy. Consumer information, organizational champions, tradeable permits, and a host of other mechanisms may need to be applied simultaneously to achieve results.

And even when the resource-threat relationship seems simple and direct, such as Amazonian deforestation, establishing effective extractive reserves may require a constellation of policy changes. Some of these changes will be active (such as identifying markets for the extracted commodities) and some will be passive or inactive (such as not giving land title or not building roads). Given the above, imagine the complexity that is likely to accompany attempts at systemic change. And systemic change cannot be controlled—only guided.

As Chapters Two and Three suggest, different intended consequences require different implementation approaches. We need to know what we are

FIGURE 3–B: Implementation Mechanisms Matching Clusters and Objectives

Consequences	Policy Clusters			
	Self-Management	Indirect Incentives	Direct Incentives	Command & Control
Less Damage	a,b,c,g,h,d	k,l	a,d,i,l,m,n	b,c,e,f,j,n
Restore Resource	a,b,c,g,h	k	a,d,i,n	b,c,e,f,j
New Behavior	a,b,c,d,g,h	k,l	a,d,i,l,m,n	b,c,e,f,j
Sequester Resource	a,b,g,h	k,n	i,n	b,c,e,f,n
Efficiency	c,g	k,l	d,l,m	b,c
Market Creation	d,g	l	d,l,m	b,c,e,f, n
Adaptation	c,d,g	k	d	e,j

KEY:
a = environmental dispute resolution
b = organizational champion
c = bureaucratic reorientation
d = laissez faire/market, which includes:
    product charges/credits; deposit-refund system; tradable permits; green labeling;
    market access control; alternative income generation; property rights
e = policy pronouncement
f = regulation
g = devolution of ownership/management
h = public trust fund
i = debt-for-nature swap
j = political exhortation and mobilization
k = publicity and public awareness
l = taxation
m = subsidy
n = trade restrictions

trying to do before we decide how to do it. But there is another dimension that further complicates the picture—we also need to know how context can act to alter the impact of implementation tactics. Thus, in addition to knowing where we want to go and what means are available to get us there, we need to scout the terrain we expect to pass through. We need a context map.

NOTES

1. CAMPFIRE is an acronym that stands for Communal Areas Management Program for Indigenous Resources.
2. I am indebted to James Nations of Conservation International for this example.
3. Their argument also has an ironic twist to it—they are inadvertently promoting a government stance that takes the risk out of land investment. By trying to reduce the effectiveness of environmental restrictions they simultaneously are coddling land investors, a government role they reject for everyone else.
4. The essence of the four elements of the natural step approach are paraphrased as: (a) substances from the earth's crust must not increase in natural settings or organisms; (b) artificial substances produced by human society must not increase in nature; (c) the biological and physical capacity of nature for production and the resulting natural diversity must not deteriorate; and (d) human needs must be met by an increasingly fair and efficient use of energy and other resources.
5. Assessments of the probability that an initiative will be sustainable need to consider four dimensions.
   Ecological sustainability—not extracting natural resources in manners that reduce their robustness, or at a higher rate than their natural replacement rates, and not altering ecosystems in ways that cause their resilience to deteriorate;
   Socio-economic sustainability—continued demand for goods and services produced by an activity, low vulnerability to market changes, and a flow of economic and social benefits well into the future;
   Financial sustainability—the adequacy of financing to support activities into the future, including capacity to handle cash-flow crises; and
   Institutional sustainability—the presence of managerial capacity, performance incentives, policy support, legal rights, rules, organizational structures and resources, and values of people, groups, and organizations engaged in the activity and its support. (See McCormick and Honadle, 1999)
6. Citizens of the United States often forget this because the United States Constitution forbids export tariffs.

4
—

A Context Map:
Variables Intervening between Policies and Consequences

In 1976, one of my consulting assignments took me to Tanzania and Ghana. The work was for the Africa Bureau of the Agency for International Development. The task was to help design a regional project, with field components in Ghana and Tanzania, that would promote sustainable development by identifying, producing, and disseminating "appropriate technologies" in the agricultural sectors of the Sub-Saharan nations of Africa.

Among the many adventures, misadventures, and meetings that occurred on that trip, two strong images remain to this day. The first image is of the origin of most of the people who were key players on the appropriate technology stage—they were expatriates. The appropriate technology movement in these countries was dominated by people of non-African descent—well-meaning outsiders, but outsiders nevertheless.

The second image is of the obsession with implements. Solutions to problems were sought through the importation of tools and their introduction into the local setting. But the setting was never the primary focus of attention. A plow from India was brought to Tanzania as an answer to low productivity. Even though the soils of India and those of Tanzania might be very different, the plow was thought to hold the key. Even though the average bullock size in India was much larger than that in Tanzania, the plow was seen as the source of salvation. Outsiders were emphasizing the importation of objects to solve problems. They saw tools but were blind to context.

Over two decades later little has changed in the field of sustainable development. Our focus has merely shifted. We have graduated from the micro-level obsession with imported physical artifacts to the macro-level obsession with imported policy prescriptions. We have learned much in the meantime, but it has not necessarily made us wiser. We simply invent new forms for the repetition of old follies.

At the turn of the century, the environmentalism in the United States suffers from a similar problem. A case in point is the disposable diaper. The present wisdom is that environmentally conscious parents use cotton diapers, not the disposables because the disposables contribute to the landfill problem of the throw-away society. The diaper problem has a blanket solution.

But when context is introduced, a single solution becomes more elusive. In the Northeastern United States the scarcity of space for landfills may argue for the cotton diaper. But in the mountain shadows of the West and the deserts of the Southwest the scarcity of water may make the frequent washing of the cotton nappy more environmentally destructive. Where you are alters the relative values of the tradeoffs between the two types of product. And when the environmental impact of cotton production is considered, then the current wisdom appears even less sapient. Context matters.

Indeed, to those entrusted with the task of turning lofty plans into changed practices, context is everything. Context seems bent on reinventing new manifestations of sabotage that make ever so sensible abstractions and grand designs seem as if they are neither sensible nor grand. Wendell Berry captured this when he wrote:

The most necessary thing in agriculture, for instance, is not to invent new technologies or methods, not to achieve "breakthroughs," but to determine what tools and methods are appropriate to specific people, places and needs, and to apply them correctly. Application (which the heroic approach ignores) is the crux, because no two farms or farmers are alike; no two fields are alike. Just the changing shape or topography of the land makes for differences of the most formidable kind. (Berry, 1981: 280)

But this does not mean that context has no underlying pattern. Different combinations of soil type, slope, climatic conditions, and vegetation in adjacent areas reveal repetitive clusters. A cultivation strategy must be custom-tailored to the time, place, and people involved in its implementation, and assessing these recurring dimensions helps to determine the strategy. The same can be said for the contexts surrounding the implementation of environmental policies—they exhibit a pattern of pitfalls and possibilities, and making that pattern explicit might help to improve policy performance.

A major purpose of this volume is to create a map of the terrain that environmental policies must negotiate. We have seen that there are different types of policies with different objectives and there are many different mechanisms that can be used to apply them. But now we need to develop a context map if we are to weigh the advantages and disadvantages of using these different approaches to achieve different objectives.

We can discern two major categories of context—the *problem context,* which defines the relationship between a resource and a threat to it; and the *social context,* which defines the web of economic, institutional, and psychological hurdles that must be overcome during the design, adoption, and implementation of an environmental policy. Each is discussed below.

PROBLEM CONTEXT

There are four major dimensions of the problem context. The first involves how connected the problem is to other factors in its surroundings. The second involves the progression of the severity of the problem. The third involves the nature of the relationship between a natural resource and the threat to it in terms of the mobility of these two elements and the directness of their interaction. And the fourth one involves the number and type of political and social boundaries crossed during the interaction between problem and threat.

Discreteness

Everything is connected to everything else. This is most likely true. But it is also useless as a guide to action. This is so for two reasons. First, humans do not have the cognitive equipment needed to deal with the immensity of this. And second, humans do not have the sheer physical power to deal with it effectively, either.

A common response to this issue is to reduce problems to discrete components and then deal with each one as a separate issue. Sometimes this works. A simple ban on some type of human activity may solve some localized problems. For example, a seasonal ban on rock climbing in an area of a park may protect a threatened bird during nesting season. But in nature, as elsewhere, this does not always work. A presence or level of activity on the part of one species may harm another one but still be needed for the health of a third population. Elephants may destroy trees but by maintaining the savanna grasslands they keep the great herds of wildebeest, gazelles, and zebra in forage and browse and thus allow the lions and cheetahs to feed on the others. The essence of the problem may lie in the connectedness of many separate elements—it may be systemic rather than discrete.

But even if it is artificial, action requires reducing problem definitions to ones that can be tackled with obtainable resources. The issue is not whether to reduce, rather it is how far to go in the reduction effort. Not far enough may make the situation unmanageable. Too far may not resolve the undesirable condition. Figuring out how to do this is art rather than science, but the more scientifically informed it is, the better the chance for a desirable outcome.

Researchers are continually finding ways that species cooperate and depend on each other through space and time. These webs of interaction are called "keystone communities." Disruption of one element can lead to the collapse of the whole collaborative and interdependent enterprise. Knowledge of this is very useful for someone contemplating policy reform targeted at an element of such a community. The magnitude of the reverberations resulting from a change in one part of the system is important.

Likewise non-species based interaction can be important, too. Water, soil, wildlife, and air are all intertwined. Changing fluxes, magnitudes, or conditions in one regime can cause effects that cross into others. Although a reductionist strategy may work in some circumstances, a holistic perspective may be needed to determine if the problem under discussion actually is in one of those circumstances. Using the wrong approach for the situation is not likely to produce success. Thus, it behooves policy reformers to determine, as well as possible, the discreteness of the problem under review before embarking on a program of solutions.

Will the solution of a particular problem produce others that either make the original situation worse or add new problems that overshadow the original one? Conversely, can the mitigation of one problem loosen multiple constraints and make the solutions to others easier to achieve? Will a sequence of efforts be needed or is a single intervention adequate? These questions help to dissect the nature of one dimension of the problem.

The problem context, then, shows a dimension parallel to one noted above—just as consequences can be targeted or systemic, so too the focal resource may possess characteristics that either make it fairly discrete or ones that make it highly interconnected with other resources and elements. Which category it falls into is, of course, a matter of degree. And, as we will discover, there is also a parallel in the social context—the strength of embedded human behavior. Interconnection and lack of discreteness often get in the way of simple solutions to problems.

But discreteness can vary through time. Indeed, the introduction of the time dimension contains another key aspect of the problem context—the roles of trends, cycles, and thresholds.

Progression

There is a time dimension to the problem context. And it is not just a matter of how fast a situation is deteriorating. The trend may be fast or slow, but it could also be part of an observable *cycle* that affects the resource. For example, long-term weather variation in East Africa can influence water availability in the short-run. If the approximate length of the weather cycle is known, then it is possible to see the relationship between the present circumstance and how it will soon be. If recent population increases preceded the entrance into a low rainfall portion of the cycle, then it may be easy to predict that the upcoming rain deficit years will be difficult to get through and to recognize that measures need to be undertaken to compensate for that fact.

Many natural resources follow cyclical patterns of waxing and waning. This is especially true for animal populations. The lemmings may be the most well-known example, but cycles of rabbit numbers followed by their lynx predators are also well-documented. Changes of this nature tend to get bumped up the food chain with changes in forage reflected in herbivore habits and these changes mirrored in predator activity with a time lag. For sedentary populations, the cycle may be registered in numbers. For mobile populations, it also may change where they go or when they go there instead of just how many of them there are.

Any attempt to deal with a problem, then, needs to take into account whether or not there is a cyclical aspect of the situation. If a problem is defined as a linear deterioration of a resource, but it is actually part of a recurring pattern, then the solution may miss the mark and exacerbate the problem when the pattern enters a new phase. Non-linear phenomena can derail the effects of policies based on linear assumptions. Understanding this dimension of the problem context is crucial. Indeed, narrow linear thinking can get us into a lot of trouble when the phenomena we confront are non-linear (Honadle, Grosse and Phumpiu, 1994; Hoare and du Toit, 1999).

The order of magnitude of the cycle is important, too. Cycles measured in decades, years, or months may be within the realm of management influence, or at least accommodation. But cycles measured in millennia, centuries, or nanoseconds fall outside the timescale of human influence. And it is our ignorance of cyclical patterns, our inability to incorporate cyclical changes into our initiatives, and their ability to overpower our efforts that make these recurrences important.

Another type of non-linear progression is crossing a *threshold*. This is a scary and particularly formidable one because it is often impossible to see it coming and when it is here it is too late. It is like the straw that breaks the camel's back—who knows which one it will be? Before crossing the threshold there was one definition of the problem. After the crossing it is another one entirely. But thresholds are not always understood. In fact, some are known, some are suspected to exist but the point of transition is unknown, and others are not even suspected.

For example: populations of animals may become unviable once genetic diversity goes below a particular level and that level might be known; the accumulation of toxins in a food chain may reach a level that causes predator reproduction to drop precipitously and that level might be known; crossing a minimum habitat size may doom a migrating species to extinction and that size might be suspected; loss of a prey may turn a predator to other food and threaten the newly hunted species beyond recovery but the new prey might not be suspected; absorption of poisons into human blood streams may reach a level where cell reproduction goes haywire and cancer results but the place where the transition from benign to cancerous happens is unknown; and many other thresholds are possible. Some are known. Many are not. And often the fear of an unknown one sneaking up on us underlies hotly contested issues.

This fear lies at the heart of two central issues today—climate change and population growth. Much of the discussion of the greenhouse effect and global warming focuses on whether or not the industrial emissions of the carbon-based societies have already passed a threshold. If we are a part of the cause of climate change, can our part be reversed or have we achieved a critical mass that puts us in an irreversible situation? Likewise, a bone of contention in the debate over human population growth is whether we have entered a demographic transition that will lead to a stabilized population or whether we have already passed a threshold that will collapse our natural support systems and lead to disaster. Which side of the threshold we find ourselves on will have a major impact on what we can do about it. Crossing a threshold redefines the nature of a problem.

The growth rate of the threat may also introduce a threshold. For example, if an invading exotic species, like Eurasian milfoil, has a population doubling time of one year, then there may be a point where the nature of the problem is transformed. If it took six years to blanket half a lake, then only one year remains before it strangles the other vegetation in the entire lake. Year seven will be a threshold.

There are many factors that can constitute thresholds. For example, laws to regulate land use that emerged from a setting with a human population density of five people per square mile may collapse when densities reach five thousand per square mile. Or, food distribution and sharing customs developed when 15 percent of a country was desert may not work in a situation where 85 percent of the nation has become desert. In both of these cases a physical threshold may have been crossed somewhere between the two extremes. On one side of the threshold things worked, but on the other side they did not. Crossing a threshold changes context.

A key question, then, is "Is this situation progressing along a linear path, or is there a cyclical or threshold component to this problem?" If, as is most likely, it is unknown, then an attempt to uncover any such possibilities would be warranted. If it can be discovered, or is already known, then any attempt to remedy the situation needs to take it into account. The nature of the problem progression is key to the problem context, and just assuming that the progression is linear is not a good way to proceed.

We need desperately to be able to escape from our limited perception of the problems we face. A focus on discreteness combined with awareness of the trajectory of the problem situation might be part of that escape plan. But the mobility of the resource and the threat to it constitute a further aspect of the problem context.

Mobility

Beginning with the resource, a key question is "Is it site-bound or mobile?" Restoring the ecological health of an immobile resource such as Lake Victoria is different from protecting the viability of a population of migratory songbirds that wing annually from the Arctic to South America and back again. The focus of attention and the activation of efforts will be very different in these two cases.

The same question applies to the threat—"Is it site-bound or mobile?" A site-bound smokestack creating acid rain on a distant lake is a far easier target to monitor than a roving fishing fleet chasing finned prey over the seas. And today, that fleet is simply the manifestation of roving capital looking for a fast return on investment—it is not a community of artisans representing a way of life dependent on the sustainable use of the fish resource. It is far more ephemeral than that, and effective measures to change behavior will reflect the difference. Indeed, intervention in capital markets, or tax laws, or commodity markets may be needed. And such interventions will seem far removed from a fish protection program. In fact, effective resource protection in the future may involve more dull and detailed work with balance sheets and computers than with romantic, and dangerous, encounters on the high seas or emotional interactions with threatened species.

When both threat and resource are fixed in place, efforts to change the interaction and monitor the changes can also be site-specific. But if either is in motion, then the solutions will be based on different types of activities. In the case of a mobile (and potentially migratory) human population, for example, forest protection might be characterized by urban development far from the forest. Both the sectoral focus and the location of the effort could be far removed from a "forestry" policy or program. Mobility can affect both the spatial relationship between resource and threat and the perception of the directness of the interaction between them. Indeed, disassembling a road network that was established to facilitate mobility may be key to resource protection. Thus, changing a resource-threat relationship may involve undoing past actions as well as redirecting future ones. Figure 4–A summarizes alternative measures implied by different resource-threat interactions.

The issue of the directness and visibility of the resource-threat interaction is also important. Although the connection between a site-bound smokestack and a site-bound lake experiencing acid rain may not pose problems of capturing mobile actors, it can still offer an example of less-than-obvious cause-effect relationships. Indeed, for many years the link between industrial emissions and dead lakes in distant locations was unknown. And often the effects of activity also are separated in time from the activity itself, thus making the causes even more difficult to determine.

This problem of indirectness is noted by Ornstein and Ehrlich when they argue that our tendency to respond to immediate threats, such as carnivores about to pounce on us or lightning striking a nearby tree, had survival value in our evolutionary past. But the world that we have made for ourselves requires responses to threats that are indirect, convoluted in their pathways and complicated in their effects. They emphasize,

Hundreds of thousands or millions of years ago, our ancestors' survival depended in large part on the ability to respond quickly to threats that were immediate, personal and palpable: threats like the sudden crack of a branch as it is about to give way or the roar of a flash flood racing down a narrow valley. Threats like the darkening of the entrance to the cavern as a giant cave bear enters. (Ornstein and Ehrlich, 1989: 8)

and then they go on to argue for the need to alter our ways of perceiving problems to encompass more pervasive, subtle, and threatening issues such as human population increase, global warming, nuclear armaments, and the evolution of biological predators such as the AIDS virus. They argue that our sensory tools are blinding us to the new threats around us and that we need

Figure 4–A: Preferred Policies for Different Resource-Threat Combinations

	Site-bound Resource	Mobile Resource
Site-bound Threat	1. Command & control	1. Command & control
	2. Stakeholder self-management	2. Direct incentives
		3. Indirect incentives
Mobile Threat	1. Command & control	1. Direct incentives
	2. Stakeholder self-management	2. Indirect incentives
	3. Direct incentives	

new ways of thinking and new categories for organizing information to help us escape from this trap.

Boundary

Treaties and conventions are mechanisms for dealing with one of the sticky problems attached to environmental policy—the tendency of problem/solution clusters to extend beyond national borders. But this is more than just a problem of nation-state boundaries. Sub-national political boundaries, ethnic group territories, public versus private ownership, and even bureaucratic jurisdictions can get in the way of effective policy implementation. Environmental threats have a habit of crossing such boundaries and making cooperative action more difficult to achieve. (Knight and Landres, 1998)

The boundary surrounding the resource may be totally unrelated to the source of the threat as well as not matching the space controlled by those interested in changing the resource-threat relationship. Indeed, the lack of conformity between ecological and political, cultural or administrative boundaries is a common occurrence worldwide. There is seldom a high correlation between natural and human-erected frontiers—watersheds, ecological zones and geological formations usually criss-cross many nations, villages and administrative districts. Thus, the lack of congruence of these various boundaries poses a problem itself—a boundary mismatch.

People typically relate to artificial areas rather than natural ones and the complexity of artificial zones can make policies ineffective. In the United States, for example, it is difficult to recognize or measure the cost of urban sprawl because school districts follow one set of boundaries, general purpose government districts follow other patterns, markets occupy other areas, commuter sheds have different outlines, and ethnic loyalties may follow still other lines. And, of course, aquifers, wind streams, watersheds, and animal migration routes respect no artificial territories. Overlaying these produces a feel for the complexity of the boundary issue. Resource threats can cross all types of boundaries—lines on maps, edges of landscapes, and fences on the ground.

Even when a designated protected area of some type is the object of interest, the threat to resources within the area often originates outside it. For example, fires that threaten mountain parks in the western United States usually start in private land below the protected areas. This is an illustration of a threat moving uphill toward a resource. More commonly, a downhill progression where land use sends silt, mine tailings, chemicals, or other environmental irritants through multiple jurisdictions, cultures, biological landscapes, and economic zones is characteristic of threats across multiple boundaries. Water currents, wind currents, and human commerce all can transport the causes of calamity.

Whether a resource-threat interaction is confined to a single entity, or whether it encompasses dual or multiple political or ownership units will influence alternatives available to implement policy. Political boundaries (those demarcating the reaches of a political state, or polity) can be especially important. Command and control obviously has a better chance of working in a single unit situation than in any of the others (civil war would be classified as non-congruence within a formal polity), and it is extremely limited in its ability to work in the commons beyond the boundaries of political control (such as the open seas or upper atmosphere). But either because our view is narrow or because the problem has reached crisis proportions, this is often the remedy of choice even for cross-polity problems.

International treaties and conventions are a common approach to situations where trade, or resource harvesting activity, or the resources themselves, cross national borders or occur in the global commons. The North American Waterfowl Management Plan, the international ban on ivory traffic, whaling conventions, and the Convention on International Trade in Endangered Species (CITES) are just a few examples.

CITES, however, contains an ability to relax command and control and thereby offer an incentive for sustainable management. The salt-water crocodile of the south Pacific is an endangered species. But there is a healthy population of these animals in Papua New Guinea (PNG). Since CITES allows exemptions from trade bans, and one has been awarded for PNG's salt-water crocodiles, the country has been rewarded for its efforts to protect this resource. It receives both international recognition and economic benefits from its policies and efforts. (And some have argued that the universal ban on ivory has failed to reward those countries in southern Africa that have done well in the management of their elephant populations and thus the long-term impact of the total ban will include negative consequences.)

The degree of match among boundaries is an aspect of context that policy makers and implementers frequently encounter. When multiple polities are involved, questions arise concerning the degree of political implementation capacity in each polity or unit, mechanisms to induce collaboration, and the sustainability of any efforts. While lack of congruence certainly imposes a major obstacle, there are tactics besides lengthy treaty negotiations or forceful annexations that can help to overcome it.

Earlier (see Chapter One) this study noted the widespread occurrence of rational self-benefitting behavior. People do generally act to reap personal rewards and they assess the tradeoffs involved in different action paths. This fits nicely into an economic view of human behavior. But there is also another type of behavior that is equally widespread. Indeed, its frequency is so great that even biologists have attempted to explain its possible genetic foundation (Wilson, 1975). It is called *altruism*.

Altruism appears when people act in ways that mainly benefit others. But the problem is how to induce it. This may be especially true in trans-boundary circumstances. When protecting a resource in one locale requires changing behavior in another, it is extremely difficult to use command and control, and the reach of other policy strategies may be equally inadequate. In such cases, altruism may offer some hope.

Fortunately, some experience exists that does indicate that a "sense of fairness" can be used to promote beneficial behavior. Although it is not trans-boundary in a national sense, it does capture the upstream-downstream nature of many trans-boundary dynamics.

The experience comes from irrigated water management schemes. An example is from Gal Oya in Sri Lanka (Uphoff, 1992). The problem was to get upstream farmers to release more water to those lower in the system. Control strategies could be subverted. The key to system success was to get voluntary compliance. This was achieved through visits to the downstream sites and meetings with the people affected. Once a face, name, and personality were associated with losses induced by water diversion, some farmers found it more difficult to do it. A sense of fairness prevailed.

But this was not achieved through lectures or "consciousness raising." It happened as social bonds were built among competing parties. Zero-sum was not erased, but socially responsible behavior was redefined through human interaction. New ties were forged. Establishing new, or building upon pre-existing, collaborative relationships across borders is a parallel to the irrigation experience. Indeed, programs to do this may be prerequisites for managing problems across polity frontiers, and such programs are invariably communication-intensive.

Synchronized donor-assisted programs in the upstream and downstream nations might be used in some cases. But outside assistance tends to be short-lived. Building lasting relationships will be key. This is complicated by political boundaries and ethnic rivalries.

Promotion of *commercial ties* would be a parallel approach to the Gal Oya experience. Building mutually beneficial relationships that create a common interest in the well-being of the resource may have staying power.

Engaging in *joint ventures* that protect the resource in one country, but concentrate value added or other foreign exchange benefits in the second country might offer possibilities. For instance, country A could export raw materials to a processing plant in country B on concessional terms. In return, country B stops an activity that threatens a resource in country A. Or, an ecotourism industry in A could purchase materials or route its tourists through B's airline or capital city. Synchronized donor investments in both components could help to begin the process. A heightened sense of fairness could provide the foundation for building mutually beneficial ties.

Creating new *cross-polity resource-based management structures*, or protection organizations, is another option. Institutions like the Great Lakes Protection Fund in the United States and the Mekong River Basin Commission in South East Asia are examples of such structures. Non-governmental organizations often follow natural boundaries, too. The Chesapeake Bay Foundation (CBF) in the United States is an example of this. Since the Chesapeake is basically the mouth of the Susquehanna River with a few other rivers, such as the Rappahannock and the Potomac, adding to the estuary, CBF must deal with the states of New York, Pennsylvania, Maryland, Virginia, and West Virginia to cover the Chesapeake catchment area. Failure to include any single state in its program can threaten the quality of the Bay.

In situations where the resource itself spans national boundaries, collaboration can be facilitated by international funding arrangements. For example, the Global Environmental Facility (GEF), established along with the UNCED initiative and placed under the aegis of the World Bank, is being used as a catalyst for international cooperation in the rejuvenation of Lake Victoria in East Africa. This large financial pool provides an incentive for collaboration among Kenya, Tanzania, and Uganda—the three nations enclosing the lake.

Another example of the GEF bringing together multiple polities is a trilateral trust fund located in Switzerland with the earnings of the trust to be used for establishment and management of adjacent parklands in Slovakia, Poland, and the Ukraine. Clearly, global or regional institutional wealth helps to overcome cross-polity implementation barriers. The source of the threat may be outside the boundary of the resource, but the solution to the problem may originate there also. And it may even promote within-boundary capacity building.

The completion of either National Environmental Action Plans (NEAPs) or National Conservation Strategies (NCSs) by June 30, 1994, was made a requirement for qualifying for access to International Development Association (IDA) support for the next replenishment. This is the loan window at the World Bank that provides concessional rates to borrowers and it is a very important funding source for the lowest income countries. The next step would be regional coordination of these NEAPs/NCSs as a qualification for the replenishments of the future. The IBRD/IDA/GEF resource pool has a potentially powerful role to play in cross-polity cooperation if it is well handled.

The match or mismatch of the area where control mechanisms work to the area occupied by resource/threat interactions will influence the implementation of an environmental policy. A trans-boundary situation is part of the problem context that must be taken into account. Likewise, the discreteness of the resource and the progression of the conditions surrounding the resource affect the nature of the problem. But there are also factors that lurk within the borders of each entity regardless of the nature of the problem context. These factors vary greatly because of the variety of human-devised settings. They constitute the social context.

SOCIAL CONTEXT

When discussing policy formulation and implementation in the developing nations of Africa, Asia, Latin America, and the islands between the continents, there are two basic social context arenas—the international arena where regional and global organizations pressure nations to sign treaties, adopt policies, and change practices; and the arena within nations where bureaucratic competition, regional jealousies, and the politics of personality reign supreme.

These will be called the macro and the micro contexts.

At the macro level, patterns of international debt, shifting terms of trade for national exports, world opinion, global climatic change and resource loss, international financial markets, multinational corporations, and a constellation of international organizations all wield influence and constrain the autonomy of nation-states. Indeed, the one/two punch of the International Monetary Fund and World Bank can greatly influence the internal policies of third world governments. Less influential, but sometimes important, environmental organizations such as the International Union for the Conservation of Nature (IUCN) or Greenpeace also can play important roles in the limiting of national autonomy.

External debt also may be significant in explaining sources of commitment to environmental policy reform. With the World Bank insisting on NEAPs or NCSs as a precondition for IDA financing, the pressure to show progress in NEAP activity is intense. But in a country like Botswana, which has the fourth lowest external debt ratio of the 101 low and middle income countries listed in the 1992 *World Development Report* (World Bank, 1992a) and which borrows close to commercial rates, the IDA stipulation carries little weight. Thus the pattern of international financial flows will affect the influence of global lobbies on national environmental efforts.

Partly due to such international influences, what appears as commitment to a treaty or regional policy on the part of a national government is sometimes little more than a ploy to gain access to international funding or lower bad publicity that threatens access to international markets. It is like an *ante* in a poker game, and it reveals little about the true intentions of national actors. When environmental policy implementation begins in earnest, international protocols may fall by the wayside to reveal intense opposition to seemingly common objectives held by international funding organizations and national bureaucracies. Micro dynamics do not always reflect macro policies because many of those policies are espoused but not followed.

One of the reasons they are not followed is that the problem addressed may have a cross-border dimension. That is, the benefits of a policy change may be reaped by the inhabitants of a neighboring or distant polity and not by the citizens of the policy-making entity. Thus the problem context can affect the social context at the micro level.

The penetration of the macro level into the micro level mixes with local-national dynamics. International organizations often possess better technical information about economic or biological trends inside a third world country than that country's own government. This gives such organizations tremendous leverage in discussions. But national politicians and bureaucrats tend to be absorbed in the politics of governing and they bend the objectives of the international organizations to fit their own agendas.

And even when there is no conscious bending of objectives, it often turns out that local perspectives are not in accord with national perspectives, that local ways of doing things were violated in the national policy dialogue, or that changing circumstances make implementation more problematic. The micro context can be full of surprises.

This is the context of policy implementation in developing countries. The micro arena is the location where environmental policy implementation occurs. But the macro influences the micro. This also holds true within the wealthier nations. Logging communities of Washington state and farming communities of rural Minnesota feel pressures resulting from state, national, and international happenings. Climate change, international migration, national concern over local resource use, international treaties, and many other factors impinge upon the autonomy of local jurisdictions. And in the United States, legal sovereignty resides only at two levels—state and national—and this sometimes weighs heavily on local actors and government units who view themselves as their own masters. The fact that their authority is only what is given by the state does not fit either their self-images or their aspirations.

Our primary concern in this study is the micro context. But, as noted above, the macro makes it's presence felt and cannot be ignored. Our task is to identify crucial aspects of the micro context that also catch the most important points of penetration by the macro. But any discussion of context encounters a sticky problem—the possible dimensions that might be included are infinite. There is no agreed upon point at which the search for contextual ingredients stops. There is no way to determine that the list catches all the important things.

The most we can do is construct a net that catches what appear to be the key opportunities exploited by reformers and the major obstacles that confront implementers. The dimensions presented here do not offer closure, either. But they do contain the following attributes:

- *First*, they all appear time and again as strong influences on implementation outcomes and strategies in a wide range of circumstances;

- *Second*, they directly relate to implementation; and

- *Third*, they offer alternative explanations for past performance that create new insights into the reasons for that performance.

The dimensions that will be presented do not constitute a true theory of context. Indeed, we have not had a complete theory of context since Fred Riggs proposed his theory of "prismatic society" in 1964.[1] But when the six constituent dimensions introduced below are linked to an assessment of how embedded they are, they comprise a conceptual framework for assessing the social context of policy implementation. They are less comprehensive and more limited than a complete theory of context, but they are also much more applied than any such theory is likely to be. Indeed, interviews and observations suggest that some policy reformers intuitively use parts of this framework. For example, key decisions often depend on assessments of unspoken elements, such as the charisma of certain actors, the balance of power among different organizations, and the openness and objectivity of information flows in the local setting. And these are all integral to an emerging picture of context.

Informational Openness/Political Culture

In 1990, I was in Papua New Guinea. During that visit a new forestry policy was being debated in the legislature. This debate, along with accusations of improprieties on the part of high officials, was carried in the newspapers. Different views were expressed in daily editions of multiple publishers. There were different parties openly declaring different perspectives on major issues.

The year before I was in Malawi. There, the newspapers were all owned, or controlled, either by the Malawi Congress Party (the only legal political party) or by Press Holdings, both of which responded directly to the president. The single radio station was government-owned. Debate here was about the trivial, and divergent views were simply not expressed for fear of retaliation.

That contrast between Malawi and Papua New Guinea was not only striking—it was central to the choice of strategy for natural resource policy implementation and it epitomized the extremes in which policies are implemented. It was context in its barest form.

The informational openness and the political culture comprise a key factor to be considered in choosing a policy reform strategy. When people say "We don't do it that way here," they often are referring to the political constraints on information flows. Indeed, in some countries of southern Africa it has been illegal to discuss or publish data on pollution levels (Chapman, 1993).

Informational openness and human rights go together. Suppression of dissent about natural resource issues is found throughout the third world (Human Rights Watch, 1991; Human Rights Watch and Natural Resources Defense Council 1992). And although open societies may occasionally, or even systematically, suppress environmental dissent, closed societies will do it more routinely and more brutally.

But technical experts working for international agencies tend to ignore the importance of human rights. They adjust to local bureaucratic styles, but they gloss over the effect of informational openness on program success. Only recently have such organizations integrated governance issues into their perspectives (World Bank, 1991).

Experience does exist, though, on the introduction of new information technologies and the contextual impact they have. For example, the use of microcomputers for district budgeting in Kenya changed the budgeting process by increasing the transparency of priority-setting and allowing all districts to view the resource levels available to others (Leonard, Cohen and Pinkney, 1983).

Experience also suggests that one of the tests of openness is effective political opposition and the threat of political turnover. If the media have a chance to bring down a regime then their existence can threaten politicians and a social commitment to openness is needed to ensure their continued operation. But when opposition is marginal, the trappings of free media are no threat (Grant and Egner, 1989) and of less consequence.

The historical trend is toward more openness, but change can be wrenching. For example, the donor-promoted referendum on a multi-party system in Malawi in 1993 led to rear guard actions and uncertainty. The march toward freer information flow was neither willing nor swift, but it may be inevitable. Satellite technology and global information networks have forever altered the politics of information.

Often the informational dimension is less dramatic than green-group demonstrations, multiparty politics, computerized budgets, environmental reporting, or freedom of the press. Sometimes it is just rules that constrict organizational operations. For example, Non-Governmental Organizations (NGOs) operate freely in Botswana, but in neighboring Zimbabwe they must be registered with the Ministry of Finance, they must hold objectives consistent with government priorities, and they are allowed to engage only in operations approved by the government (OTA, 1988).

The key reason why informational openness is so important is that it determines risks encountered by proponents of policy change. The more open the system, the more the risks can be buffered and protected by the media and allies (LaMay and Dennis, 1991). The more closed the system, the greater the difficulty of reading risk and promoting change.

Even relatively open societies will suppress information when it threatens powerful interests. For example, the United States has freedom of information laws and protections for whistle blowers. But it also exhibits cases where bureaucratic, industrial and political actors have persecuted those who try to shine a light on either the impact of policies that fly in the face of good science or the subterfuge that accompanies distorted policy implementation and captured agencies (Wilkinson, 1998).

These pervasive issues of human rights, informational openness, and political culture limit options for promoting participation, for setting incentive systems and for going beyond command and control approaches. But eventually strategy alternatives narrow to the point where the choice of implementing organization becomes central. And to make that decision it is necessary to understand the interorganizational power balance.

Interorganizational Power Balance

The balance of resources and agenda among government ministries and other organizations will shape the operation of a policy reform effort and it is necessary to assess that balance when designing the reform. For example, in Thailand, where the forests are so depleted that logging is now illegal, the ministry of forests is charged with protection of forests and national parks. At the same time, there is a conflict with a ministry that would seem to be a natural ally—the Ministry of Tourism.

Rather than promoting ecotourism, the Ministry of Tourism sees its mission as championing the construction of roads and artificial lakes to attract the Asian tourist trade and capture foreign currency. This may be an accurate reflection of present demand for leisure opportunities among the wealthy classes of Asia, but it directly pits one ministerial agenda against another. Relative budgetary strength and access to sympathetic and powerful actors, such as the military, will be important determinants of which perspective dominates. Any attempt at forest protection will need to recognize this.

In other countries the interorganizational dimension takes a similar, but not identical, form. Relative budgets reveal much about true priorities. When a ministry devotes most of its budget to a production division, and that division of one department has a budget that greatly overshadows the total budget of an entire regulatory agency, such as Papua New Guinea's Department of Environment and Conservation, then the task of the second organization is made much more difficult. This particular scenario is played out in many tropical countries as well as in some major nations of the northern hemisphere.

Overlapping jurisdictions among the various ministries charged with resource management also can complicate the picture. In some countries mangrove and other coastal forests come under the authority of a different ministry than upland forests. Or contradictions among the roles of local governments and departments of agriculture, forestry, and public works can delay the identification of authority to stop questionable practices until it is too late and the forest is gone.

Unclear division of responsibility and authority serves the interests of exploiters in cases where the strongest leadership resides in the organizations sympathetic to the mining of the resource. Where the stronger leader heads an organization aiming at protection and sustainable use, then the protection agenda might temporarily prevail as a result of fuzzy jurisdictions. Protective action can precede sorting out the legalities. But each case must be assessed according to local circumstances.

Other system-wide practices such as reimbursement procedures for civil servant expenses, or monthly payment systems for all ministries can greatly constrain the activities of line ministries but not parastatals.[2] Sometimes important factors come from across national boundaries. For example, Thailand has banned logging but not the transport of logs. So timber from Laos and Myanmar makes its way through Thailand—deforestation has been exported.

Organizational characteristics of the local environment and an understanding of the balance of power among organizations, are crucial for choosing policy implementation strategies in a wide range of settings (Clarke and McCool, 1985; Gallagher, 1991; Hurst, 1990; Ledec, 1985; Lindenberg and Crosby, 1981). In fact, one observer has argued that without first sorting out these institutional problems, donor investments in natural resource policy reforms are little more than money down the drain (Buckman, 1987). So this must be considered a key dimension of context.

Indeed, a policy implementation strategy must be designed to cope with interorganizational dynamics. Understanding the tendency of organizations to compete (Hough, 1994) and to try to transfer risks to others (Crocker and Shogren, 1994) is essential when devising implementation approaches. Both competitive and cooperative possibilities must be understood, and strategies must incorporate measures to address sources of opposition and power imbalances (Honadle and Cooper, 1989).

This power balance issue ranks among the primary sources of contention during the design of organizations to implement NEAPs and NCSs. Organizational placement questions and debates about "executive" versus "advisory" or "coordinating" functions and authority all reflect this (Brinkerhoff and Gage, 1993; Brinkerhoff and Yaeger, 1993; Dorm-Adzobu, 1995; Gustafson and Clifford, 1994; Honadle, 1994). Indeed, the power balance affects whose definition of an environmental "problem" will hold sway. Conversely, the relative leadership merits of an organization are influenced by the way the environmental problem is defined. And an important part of this definition flows from the salience of the policy issue and the speed of the deterioration of the natural resource at risk.

Salience

Just as necessity gives birth to invention, so too opportunities spring forth from crises. Debt-for-nature swaps, for instance, would not have appeared except for the staggering debt burdens facing the borrower nations in the late 1970s and early 1980s (Miller, 1991). When mere difficulties turn into crises, then major changes are possible. People are more likely to seriously consider radical change when faced with a need for immediate action on a critical issue. This dual sense of importance and immediacy is the essence of salience. A major problem combined with an urgent need to solve it results in a salient situation.

This is where the importance of individual leadership enters the picture. The significance of key actors cannot be ignored (Grindle and Thomas, 1991; Brooke, 1993; Leonard, 1991). But for such leadership to emerge, there must be recognition of the salience of the issue. Someone must define the situation in such a way that the risks attached to challenging the status quo are offset by the perceived gains. The motivation is not important—altruism, expanding a personal power base, protecting investments, weakening a rival individual or group, religious beliefs, or professional interest may all be motivating factors appearing in many mixes. But the human spark emerges from an appreciation of the salience of the situation and the determination to seize the moment.

That moment is often created by the visible, measurable and major deterioration of a physical resource. For example, in Papua New Guinea in 1990 the damage caused by mine tailings reached such proportions that it could no longer be ignored; in India, the incident at Bhopal brought immediate attention because it was major, dramatic, and undeniable; in the Sahel region of Africa, the advances of the desert in the 1970s displaced thousands of people and created immediate problems; on the Indian subcontinent the conflict over limited water supplies has exacerbated regional political conflict. Such occurrences are the crucible of change.

Likewise, financial resource depletion can bring the problem to the fore. International debt crises can grab the attention of politicians, or rapidly falling prices for a country's major export can induce rapid response. The overnight appearance of political refugees can tax a country's support systems and internal population growth can increase strains on natural resources.

When salience is lacking, however, methods imported from a place where it was present might not work. For example, voluntary approaches that work in the Netherlands may be ignored where there is no equivalent to a wall of water continually poised to overwhelm the countryside.

Recognizing this, many environmentalists try to induce a sense of salience to muster support for their programs and agendas. Although a sense of salience generates political, bureaucratic or community support, the paradox is that such a sense may not be founded on fact. For example, first world support for tropical rainforest protection has been intensified by referring to the forest as "the lungs of the earth" and implying that they export oxygen. But, in fact, they generally are in oxygen balance (Hecht and Cockburn, 1990). Likewise, forest protection in Central America has been mobilized based on erroneous interpretations of climatic factors. And reactions to irradiated vegetables may act more to line the coffers of advocacy groups than to protect consumers (Fumento, 1993). Thus salience may be based on either fact or fiction.

Even so, increasing rates of natural resource exhaustion colliding with soaring demand for those resources as a result of human population surges is likely to increase real salience. The perceived intensity of the situation may provide implementation opportunities. As the age of incrementalism is replaced by cascading crises, more radical solutions become palatable and the range of acceptable options is increased.

And there is evidence that this has occurred. For example, the level of general development assistance flowing through international organizations in the mid-1990s was no greater than the 1990 level even though the dollar figures rose substantially. This was because the real increases over four years were gobbled up by the rising percentage of Official Development Assistance going to emergency relief (UNDP, 1994). Crises are claiming more and more of our resources.

This is important for environmental policy implementation because at the close of the twentieth century most disasters involve environmental causes or consequences (Myers, 1994). And when international environmental crises reach disaster proportions, it is easier to get nations to act. Likewise, urban clean-up in the United States is easier when the situation becomes intolerable. In the words of Annie Young, an urban environmentalist in Minneapolis, Minnesota, "It's in Pittsburgh and Detroit and Chatanooga where progress is made—we only act when our backs are against the wall." High salience is a common ingredient in the formula for action throughout the globe. It is preferable, however, for it to rise before a threshold is reached, not after.

But even when salience is high, the choice of processes can affect implementation. The chances for effective performance are raised when management approaches are compatible with local cultural practices.

Process Requirements of the Culture(s)

Cultures value some ways of doing things more than other ways. And they develop signals to clarify situations and inform actors of the appropriateness of their actions. But when actions are based on approved processes in other cultures, or when signals are murky, then joint actions may disintegrate into individual efforts pulling in opposite directions.

Some activities are common to many cultures but they function differently depending upon where they are. Meetings are an example of this phenomenon. In some cultures, meetings can be confrontational venues for opposing parties. In other cultures, some forms of meeting limit interaction to polite opinion-presentation. Botswana's *kgotla* represents this second model. Here, consensus is built through seemingly infinite rounds of group consultation. Likewise, many cultures use public or formal meetings to make decisions or to gather data to be used in making decisions. This is often the case with organizations with their roots in the industrial democracies of the northern hemisphere. But, in many nations of Sub-Saharan Africa, meetings have traditionally served another function—they are used to announce decisions. Consultations among community members precede the meeting, which heralds the end of a process, not the process itself. Consensus is built prior to the gathering, which declares a formal end to the deliberations. A similar meeting in other cultures could involve debate and a vote. But either approach used in the wrong place would evoke responses of distrust, confusion, and low commitment. Process requirements must be satisfied.

Policy implementation does not require the dissection of all aspects of culture. We do not need to know everything to be successful. In fact, this has been captured by the term "optimal ignorance." That is, since we cannot know everything pertaining to an issue, the key is to know what is necessary but not to waste scarce resources learning more than is needed to deal with the issue.

Although different societies may have very different tolerances for nature, wildlife, and wilderness (Decker and Purdey, 1988), these characteristics are only some among the many factors that may influence environmental policy. And many may be insignificant. But process values and preferences are important because if policy implementation is based on wrong assumptions about these elements it may generate resistance and falter at times of stress. Indeed, avoiding the issue of process values bolsters espoused policies and lowers the chances that they will become policies in use.

Mamadou Dia of the World Bank has identified some key process elements that he thinks explain some of the failings of development management in Sub-Saharan Africa over the last three decades (Dia, 1991; 1992). One of these involves contract enforcement. Anyone who has observed African traditional court proceedings or the rulings of village elders on a conflict knows that the presence of the public is important—it seals the decision by making those present witnesses to the ruling. But it goes further than this in many traditional societies—it brings the observers in as active enforcers of the contract, agreement or ruling. Indeed, without community commitment to upholding the pact, it is not valid.

To Dia, this is a partial explanation for the failure of some African governments to uphold their parts of negotiated agreements concerning policies and projects. No third party is bound to engage in upholding the pacts between governments and donors, and thus they do not have high priority or legitimacy. This failure to observe process requirements contributes to a decline in management performance. The rules of the West African "palaver hut" and continent-wide community contract enforcement have been broken.

When rules are broken, conflict may ensue. But different cultural groups may use different processes and institutional mechanisms to mediate or resolve the conflict. The coherence of a community, the range of religious and ethnic groups comprising it, the distribution of people of different ages within it, and recent changes in location or composition can all influence process preference. And this will affect policy arenas as diverse as criminal justice, agricultural extension, poverty alleviation, banking and business, and natural resource protection.

In the state of Minnesota, USA, for example, recent migration of Hmong people from Laos has added a new element to social dynamics. Different views of marriage, hunting, fishing, and dispute resolution have led to misunderstanding and the need to deal with a new social structure, communication network and process for dispute resolution. Indeed, the Hmong reliance on traditional practice brought from Laos created a parallel social structure alongside the other elements of Minnesota society.

This also happened during the past two decades when Minnesotans found themselves facing an increasing number of people migrating to Minnesota from other states and carrying with them different dispute resolution traditions. Traditional Minnesotans avoid the use of the courts as much as possible. But many other Americans do not. Moreover, a growing population adds not only variety but numbers, also. As density of human populations increases, the worldwide tendency is greater reliance on formal conflict management institutions such as the courts and the police. And the growth of the suburbs and the commuter shed places new demands on rural institutions—demands they often do not understand and are ill-equipped to handle.

A contextual factor important for policy implementation, then, is the process requirements of the cultures in a problem area. Indeed, the "s" in cultures is important, also. High diversity within a any geographic unit, along with the mobility of different ethnic groups and changes in the density and scale of people, may complicate implementation.

Scale, Space, and Infrastructure

Small, poor countries are sometimes at the mercy of international organizations that have resources and policy leverage that overwhelm the supposedly sovereign nation states. And even field-level operations in remote areas can feel the pressure from the global agencies.

In the late 1960s I worked as an agricultural extension officer in a remote area of Malawi. Even though it was cut off by road from the rest of the country about two months each year, the Karonga lakeshore plain was a fertile area that caught the eye of international agencies. It was remote but attractive.

But the capacity of the agricultural field office in Karonga District was limited. Geographic isolation, poverty and the small size of the Malawi government all contributed to a situation characterized by poor roads, poor communications, limited human resources, and even more overtaxed physical resources. In fact, the agricultural office had few vehicles that were serviceable at any one time. When World Bank experts were expected to come calling, the Divisional Field Officer would ask me to take the two good Land Rovers to a very remote part of the district so that the visiting dignitaries would be forced to bring other vehicles. Resources were so scarce that even subterfuge was used to protect them. Scale, space and infrastructure made a difference.

Geography, resource endowments and financial wealth, and institutional capacity all affect the viability of alternative tactics for inducing policy change. Even size alone affects the nature of policy implementation. Large societies have the advantages of:

- a critical mass of organizational and human resources and a corresponding low cost per capita of policy implementation; and

- the resilience associated with geographical and economic diversity.

But they also face corresponding disadvantages such as:

- the potential for policies to be appropriate in some sections of the country, but to not reflect local realities in other sections;

- the communication and transportation difficulties associated with large distances; and

- the political difficulties associated with ethnic and regional diversity.

Small-scale and island nations, however, face a different set of obstacles and opportunities. Indeed, they are the mirror image of the large-state situation (Baker, 1992). Each advantage held by a large society is one missing in a small place, but each disadvantage of size is not necessarily an advantage of smallness. Small may be beautiful, but it can be very difficult, too. These factors are important contextual considerations for devising a strategy to implement environmental policy. Mundane as they seem, failure to examine them when choosing policy options can create unnecessary problems.

The spatial distribution of a problem, or of actors creating a problem, is also important. Command and control is far easier to implement when a contiguous reserve, or a defined corridor, or a concentration of smokestacks provide a focal point. But when millions of scattered households or businesses are the actors creating a problem, then incentive-based and self-policing policies are more likely to be implemented successfully. Just as it is difficult to deliver services to dispersed rural populations (Honadle, 1983), it is also difficult to deflect a dispersed assault on natural resources.

Physical infrastructure does not refer just to harbors, airports, bridges, roads or buildings. It also includes communication networks. The arrival of the global electronic village has obscured the fact that there are ghettos of information blackout within that village. Indeed, in many third world countries it is easier to phone, fax, e-mail, or otherwise use satellite technology to communicate with other parts of the world than it is to interact with domestic locations. Local phone systems are captives of physical decay, human error and the whim of the weather, whereas direct satellite uplinks and downlinks can free international organizations from many of the vulnerabilities of the older technologies.

Infrastructure also has an organizational dimension—it is not just physical. In some settings there is a paucity of organizational alternatives to choose from, while in others there is a very dense population of local organizations. Organizational density not only allows choice, it also changes local social dynamics by influencing competition versus monopoly and the intensity of interactions.

In some places, such as Italy, the historical persistence of organizational density explains much in terms of relative quality of life (Putnam, 1993). And in Africa, the density of voluntary organizations makes Kenya a rich source of success stories about agricultural intensification and "green" initiatives at the grass roots (Honadle, Grosse and Phumpiu, 1994). Unfortunately, there is a tendency to generalize from one type of setting to another and to act as if context makes no difference. In fact, an attempt to transfer a project strategy from organizationally dense Zimbabwe to its organizationally sparse neighbor, Botswana, required major adjustments because the organization prerequisites were not in place (Odell, and others, 1993). Indeed, organizational density and intensity must be included in meaningful definitions of infrastructure (Cernea, 1992) because infrastructure includes institutional presence and strength, and this affects policy options (Robinson, Redford and Bennett, 1999).

And institutional strength is built upon human resources. Education, technical proficiency, creative thinking, and even literacy and numeracy are essential building blocks for organizational capacity. They do not automatically equal capacity, but they are key parts of it. These factors must be considered when developing policy implementation strategies. Effective implementation may require efforts to create the conditions that enable resource policies to work—markets, fiat, and local resource management all need supporting infrastructure.

In some settings the policy-project sequence may be backward—policies will remain unworkable until the investments are made to install the infrastructural prerequisites. In other settings, however, the presence of infrastructure can make policy implementation more difficult. The scourge of protected areas is the road—access leads to population inflows, land speculation, and the attendant environmental degradation. Nevertheless, the fact remains that the presence, absence, type, and condition of infrastructure all affect policy implementation.

Different magnitudes of scale, space, and infrastructure, then, can be expected to exert profoundly different pressures on policy implementers. And no factors are static—changes in both direction and speed will alter the context of policy implementation.

Resource Decision System

A tidier title for this dimension would be "resource regime." But unfortunately this term has been used to describe general categories of ownership and control of natural resources (Young, 1981; Young, 1982) or distinctions among different degrees of common property-related management arrangements (Bromley and Cernea, 1989). And we are not interested in general typologies here. Rather, the idea of a resource decision system is based on the question "Who makes decisions about the use of this resource in this place at this time?" It is a very specific question related to the policy consequences sought in a particular setting. An example from Lesotho might help clarify the point.

A donor-assisted land conservation and range management project encountered difficulty in understanding how decisions were made concerning the movement of livestock from one range to another at different times of the year. Although a traditional village authority structure existed, two factors complicated the issue. First, traditional decisions about livestock were entrusted to the men who were the owners of the animals. But in rural Lesotho most households were headed by women. The men were absent, working in South Africa. Such absence made it difficult for rapid response to weather conditions or other occurrences. A long and time consuming decision chain would threaten the program. Second, it was impossible to enter a village, observe the animals in the compound or in mountain pastures reserved for that village, and then understand the movement or control of the livestock population. That was because of a practice called *mafisa*.

Mafisa was a system of loaning chattel among individuals, families, and communities. Hoes, sheep, cattle, ponies, carts, and other possessions could be loaned over long periods from one unit to another. And then they could be reclaimed. Thus the items observed in one village today might turn up in another tomorrow and the contractual understandings guiding animal husbandry might suddenly change (Murray, 1981). Since the program emphasized controlling livestock numbers and movement to lower soil erosion and grassland deterioration, the complexity and unclear understanding of the resource decision system made implementation more difficult. Indeed, it was a key aspect of resource management and it perplexed the project team.

If a policy reform is based on an erroneous understanding of the operating resource decision system, the consequences can be quite unintended. Incentives may not work because the wrong people are targeted, local management may be based on the wrong units, or command and control may be unable to counter the complex, mutable and mysterious distribution of decision-making authority. The key is to comprehend what system is actually operative in the arena where the policy will apply (Honadle, 1982a).

And this applies to bureaucratic arenas as well as village settings. The *mafisa* system noted above had a parallel in the Lesotho government—it was not possible to observe a vehicle from a specific ministry in a remote area and conclude it was being used for the work of that ministry. The petrol may be supplied by one organization, the vehicle by another, and the passengers may represent multiple agencies. In another case, personnel assignments in Indian bureaucracy may result from a market for good postings based on the graft potential of the post (Wade, 1985). Not understanding this resource decision system could produce policy recommendations leading to consequences that were utterly unintended. In fact, a concern with organizational options for social forestry programs is really a focus on fitting interventions into the local decision system (Cernea, 1985). And the need to fit the situation also holds true for the private sector.

If a policy targets behavior surrounding the disposal of municipal waste, for example, then actual decision making in the jurisdiction will be key—and public sector/private sector categories may not yield insights into what really happens. Indeed, the key is to identify the central stakeholders—those who "have access to resources that are needed to carry out an activity, or have resources that can be mobilized to prevent the activity from being performed" (Honadle and Cooper, 1989: 1532). Only then are the policy and the implementation process likely to lead to the desired results.

Although general categories of resource regimes have been identified, such categories are of little help to policy implementers. The variations on a theme seem infinite, and it is those variations that spell the difference between success and failure. Take the case of property rights. Much discussion of this issue promotes generalizations and obscures the variations. Sometimes this takes the form of a blanket call for "clear" or "secure" property rights (Panayotou, 1993). Other times it takes the form of inferring that rules found in one location apply elsewhere, without examining if, indeed, that is the case.

For example, the point has been made that trees in India are state property. Thus, when a private landholder plants a tree he can create a resource owned by the state rather than himself. Tree planting, then, can take control of the land away from the rural dweller. This generates opportunities for corruption as forestry officials look the other way or it provides disincentives for planting trees to ease the fuel wood shortage. And similar things do happen elsewhere, such as state assumption of tree ownership by the current regime in Ethiopia. But this specific practice in two locations has evolved into an assumption that the same thing happens universally.

A conversation with a donor official about African land tenure was riddled with questionable generalizations based on these Indian and Ethiopian examples. Among the Nyakyusa and Ngonde people of Northern Malawi and Western Tanzania, for example, this is not the rule governing tree ownership. Indeed, to know what the situation is, one must also know whether the trees in question are banana, cashew, or other species, whether they occupy homestead or field sites, and whether the location involved is lakeshore, upland, or in volcanic craters (Gulliver, 1958; Wilson & Wilson, 1945; Honadle, 1980). In some situations, the trees remain the property of the one who planted them even years after the planter has moved away and the use of the land has been assigned to someone else by the Village Headman. In other cases, the trees go with the land. In no cases do privately planted trees automatically become state property. But to this bureaucrat, the Nyakusa/Ngonde practice was simply an aberration. The attention given to the finding in India had propelled it into a norm. But it is not. The cases of India and Ethiopia are just that—isolated cases and not a planetary practice. What matters for environmental policy implementation is the resource decision system operating in the micro region where implementation occurs (Hitchcock, 1987).

Wildlife "ownership" also varies from place to place. In much of Europe and Australia and New Zealand wildlife belongs to the owner of the land on which it is found. If a deer moves from ranch to ranch it transfers owners (unless it is royal game). In the United States, however, wild game belongs to the state—not the proprietor of the land. In the state of Minnesota this also extends to surface water—the landowner owns the ground under the water, but the state regulates activity in and on the water itself. And in some places mineral rights are automatically attached to a deed for land, whereas in other places they are not. The right to exploit minerals may reside with the state, not the landowner. Again, the particular rules surrounding a particular resource are key.

But generalizations are popular because they support a desired characteristic in donor projects—replicability. The push for replicability encourages the acceptance of generalizations because it fits into the drive for standardization and it enhances the illusion of organizational learning. An example comes from the Natural Resources Management (NRM) project in Botswana. This project was designed to replicate aspects of the CAMPFIRE experience in neighboring Zimbabwe. But the resource decision system was different in Botswana, the effects were not the same as those in Zimbabwe (Odell and others, 1993), and the strategy was adjusted to accommodate those differences. An approach using the existing local government and village structure fit Zimbabwe's social and political circumstances, but a different tack was needed to fit Botswana. The chosen course was to establish new legal personae outside the existing structure. These new entities were called "community trusts" and they reflected a policy-setting conducive to private sector based initiatives. They also reflected the need to take into account low population densities and multiple villages within wildlife management units. Even close neighbors exhibit diverse micro-environments and a sure recipe for failure is to replicate project characteristics without understanding the interplay between those characteristics and context.

This admonition even applies to CAMPFIRE experience within Zimbabwe. During a field investigation there in 1995, I addressed this issue by questioning close observers of the CAMPFIRE scene about reasons for the differences in performance among different CAMPFIRE locations. Their answers clearly suggested that the presence of the following five conditions promoted success while their absence threatened it:

- *Supportive Community Dynamics*—ethnic homogeneity and cohesiveness within the community combined with a high-density settlement pattern and low migration into the area;

- *Appropriate Resource/Settlement Relationship*—a high-value resource matched with low population size;

- *Matched Transportation/Resource Fit*—a large distance from an all season road for wildlife resources and proximity to an all season road for non-wildlife resources;

- *Appropriate Interorganizational Network*—an effective network of support institutions with adequate density and capacity; and

- *Local Leadership with Authority and Capacity*—able individuals operating within an appropriate policy setting, legal framework and organizational structure.

Thus, the match between the resource to be managed and elements of the local decision system explained performance differences within Zimbabwe. This perspective also worked extremely well in a review of community-based natural resource management in Botswana four years later (McCormick and Honadle, 1999).

These differences also were related to rates of change. For example, where migration into the area was on the increase, anticipated problems were greater than where it was declining. And the speed of change was also important in both the management process and perceptions of the problem. Changing contexts can be like the frog in the heated water—if the temperature is gradually increased the frog is unaware of the danger and will not jump out, whereas a quick rise prompts a response.

Differences between espoused and practiced decision systems can also negate the effects of policy. In Kenya, registration of land did not necessarily reflect the control of the land. In Malawi, Tanzania, and Zambia, very different practices were intermingled and in flux (Honadle, 1980). In Papua New Guinea a unique land management system evolved (James, 1985). In fact, resource decision systems continue to evolve. Just as the stability and "balance of nature" is a myth (Botkin, 1990; Drury, 1998), so too the view of a constant traditional society is a myth—all are changing. The decision systems encountered during the colonial period in Africa were evolving due to population pressure, migration and climatic cycles. They continued to change as colonial regimes attempted to freeze the traditional systems and superimpose imported legal systems and market linkages. And further evolution occurred as newly independent, but weak, states attempted to capture them, enclose them, and fold them into a new polity. In Southern Africa, Zimbabwe, Namibia, and South Africa are presently embarking on early stages of the process encountered elsewhere in Africa decades ago.

And change continues. Human mobility in the 1990s is far greater than it was one hundred years ago. Today in West Africa migration and transient job seekers often result in large percentages of local populations not adhering to local rules—they consider themselves temporary residents without a long term stake in wisely using or conserving resources. This was also reflected in the CAMPFIRE experience noted above. In other places, refugee numbers have swelled to large proportions of the total national population and this introduces still other resource management approaches. The aftermath of civil war in places like Cambodia or Rwanda or Mozambique can be an inability to agree on what rules are legitimate. The result of all of these situations is systems in flux and chaos.

When resource decision systems are in flux they often contain pressure points where policies can be targeted to guide the direction of change. Demographic pressure, for example, can cause great changes in land tenure systems (Gulliver, 1958). When change is in progress, policy interventions can channel that change. But it is necessary to understand the direction and pace of change and the reasons for it. Otherwise, policy remedies may exacerbate system failings (Panayotou, 1993). The key is to understand the matrix of social relationships that guide decisions about the use of natural resources and resource products. This is a paramount element of context.

The six dimensions of context introduced above will all be important when developing an environmental policy implementation process. The relative importance may vary from place to place, and any one can be the most daunting factor in a particular setting. But some combination of these elements is likely to shape the implementation terrain encountered by an environmental policy.

For example, in one country the combination of resource decision system and interorganizational power balance might be the overriding factors. Indeed, in the Lesotho Land Conservation and Range Management Project, a combination of traditional resource control and the project alliance with the local courts was key. But elsewhere the deciding elements might be informational openness, the salience of the situation and the process requirements of the culture, such as was found in an attempt to develop environmental indicators in Papua New Guinea (Young and Ryan, 1994). Such possible differences are displayed in Figure 4–B.

The question remains, however, of how tightly glued together these elements are, how central they are to a social system, and how immutable they are. That is the question of embeddedness.

EMBEDDEDNESS

The centrality of a resource to the local lifestyle can either complicate or simplify efforts at policy change. In 1990, as a recent migrant to the state of Minnesota from Maryland, I was struck by the contrast between local perceptions of the Mississippi River to Minnesotans and those of the Chesa-

Figure 4–B
Contextual Dimensions: Importance Varies

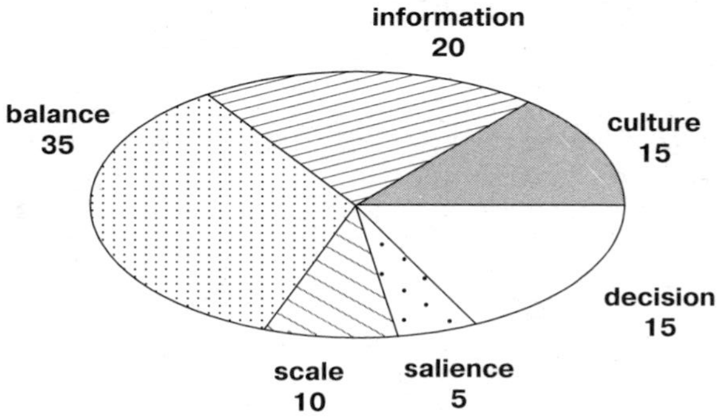

COUNTRY A

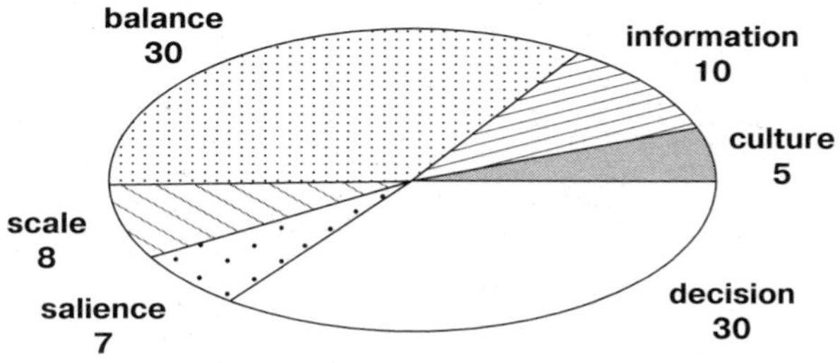

culture 15
decision 15
salience 5
scale 10
information 20
balance 35

COUNTRY B

culture 5
information 10
decision 30
salience 7
scale 8
balance 30

peake Bay to Marylanders. Indeed, like the Chesapeake, the Mississippi was in trouble. Toxic contamination, sedimentation, oxygen depletion, loss of wildlife habitat adjacent to the river, and disruption of flow were all among the signs of that trouble.

But the signs commanded little attention except during the periodic flooding or drought that forcefully injected the river into people's lives. The human settlements, industries, agriculture and commercial actors surrounding the river were conditioned to view the river system as a waste sink to absorb runoff, to provide a transportation corridor and to wash away the detritus of society. There was no tradition comparable to the oyster men of the Chesapeake Bay who lived by the pulse of the bay and suffered immediate economic and social hardship when the health of the estuary declined. There was no central set of commercial actors, such as the Chesapeake's crab and fish suppliers, hunting and fishing guides, and tourist industry, that depended upon the condition of the natural endowment to provide employment and investment (Horton, 1987; Peffer, 1979; Warner, 1976).

Those who suffered from the Mississippi's decline were largely the disadvantaged and powerless who lived far downstream and whose voices went unheard. The great majority of people measured the Mississippi not by what it produced, but by what it absorbed. Human commerce in the upper Mississippi River Basin was along the banks and surface of the river—it was not integral to its health.

This represents a dimension of context that is crucial in the development of a strategy for implementing environmental policies—the degree to which a focal resource or an intended behavior change is central or peripheral to a society. Indeed, the way a web of social structures acts to impede individual economic actions must be understood if we are to be able to predict the impact of alternative policy implementation strategies (Granovetter, 1985). It is a question of embeddedness.

Embeddedness is not just another factor to address as a single issue, but rather it refers to the connectedness among the dimensions and the amount of systemic change required to allow a policy to be implemented. It is qualitatively different from the dimensions discussed above.

An illustration from the United States demonstrates the idea of embeddedness. When the Endangered Species Act was invoked to save the habitat of the Northern Spotted Owl by halting timber cutting in the old-growth forest of the Pacific Northwest, the issue was commonly cast as "owls versus jobs." Even though the percentage of regional employment based in the timber industry had been steadily declining for many years, still this simple vision of a single trade-off dominated discussion and reporting. But the way that the logging activity permeated the society received no publicity. In fact, an area of activity central to rural life would be greatly affected—the public school systems.

In the United States a major source of revenue for supporting public schools comes from locally assessed and locally collected real estate taxes. In many counties of the Pacific Northwest states of Washington and Oregon, however, a very high percentage of the land is federally owned. This reduces the tax base considerably. To solve this problem, many years ago a policy was instituted that allowed a fee to be assessed on each log cut on federal land and for the monies generated from that assessment to be transferred to local school board coffers.[3] The practice is called PILT, for "payment in lieu of taxes." With reduced timber cutting on federal land, the financial base for the local public school system was threatened.

This was not just a threat to education, it was also a direct attack on local communities' sense of identity. In this part of the country, from the plains states to the far west, attempts to consolidate small-scale schools into a more cost-efficient system have encountered strong resistance because local sporting rivalries between schools is a major factor in community identity and definition. Indeed, communities will go all out in the fight to save their schools and their post offices. Sports scores and zip codes are numbers that define a place in the cosmos. These two items are central to local identity.

Thus, a recommended policy change encountered an embedded situation. What seemed, at first glance, to be a discrete and highly targeted reform actually turned out to threaten change at the very core of rural society. Both financial resources and social values were undercut.

Resource Dependency

An embedded situation is not an argument for not reforming policy. But it does provide evidence of the need to identify connected phenomena and to develop strategies that loosen those connections that threaten reform. In the example above, the educational finance system was the creation of human imagination and a new system could likewise be devised through human ingenuity. But embedded situations must be recognized because the transition planning required will be much, much more than in simpler situations. Likewise, implementing the transition will require a powerful process that includes provision for functional equivalents of lost items, including resources.

An agrarian reform effort in the Philippines encountered just such a lost resource. Under the traditional land tenure system in Central Luzon a share-cropper could count on support from the landowner when the sharecropper's house was destroyed by a typhoon (in many places an event that occurred about every five years). The patron provided food and shelter for the family of the sharecropper and assisted in the rebuilding of the house. But land re-form substituted freehold title for the share hold arrangement. This meant the patron-client relationship was dissolved and the rural poor lost this help. A certificate of title did not offer a functional substitute for emergency assis-tance. (Honadle and VanSant, 1985)

At the very beginning of the reform there was resistance by patrons, but sharecroppers were enthusiastic. As time went on, however, the sharecrop-pers had doubts about the benefits of participating. Typhoons revealed the vulnerabilities they encountered when they took on the mantle of ownership. New enrollment dropped. Resistance to the reform effort could have been lowered if temporary assistance arrangements were built into the program and if new title holders were given help in developing mutual support ar-rangements to take the place of those that were lost. Overcoming embeddedness requires both recognizing its existence and planning substitute relationships.

When there is a reliance on resources generated by environmentally dam-aging activities, reform can be more problematic. For example, the Govern-ment of Papua New Guinea's dependence on tax revenues from mining makes it hard to countenance policy options that endanger that flow of funds. Like-wise, diamond mining is the Republic of Botswana's major source of govern-ment income and policies are not likely to endanger that flow of funds, either.

Sometimes, however, situations are less embedded than people assume. Seizing moments of opportunity can lead to much more radical changes than were previously thought possible. In the Gambia, for example, a sweeping set of reforms, called the Economic Recovery Program of 1985, encountered little political opposition and was implemented just before a national election. A combination of citizen concerns for national sovereignty (as opposed to short-term self interest), an ability to absorb displaced civil servants into the labor force, and a fortuitous drop in world rice prices (softening the short term impact by reducing food costs), cushioning international aid flows, and a combination of a secure political regime and a democratic tradition all con-tributed to a setting that accepted radical change (Radelet, 1992; McPherson and Radelet, 1992). Indeed, the climate for change was partly used and partly created by political strategists. And embeddedness can also create vulnerabil-ity. For example, a Greenpeace threat to boycott the diamond market over threats to the Okavango Delta got the attention of the Government of Botswana very quickly.

As the discussion above suggests, there are three aspects of embeddedness. The first is *resource dependency*. The more a family, community, society, country, or organization is dependent upon revenue or services from a particular source, the more difficult it will be to engage in a reform effort that reduces or threatens those resource flows. In fact, high dependency on a single resource is often accompanied by a history of policies and decisions that have exacerbated the lack of diversity in the economy or revenue system. This is the political economy side of embeddedness.

Psychological Dependency

The second aspect of embeddedness is *psychological dependency*. The greater the link to personal or group identity and sense of self-worth that is engendered by a relationship, the greater the difficulty of introducing reforms to change that relationship. Social mechanisms often evolve to reinforce the dependency, and an attack on the source of identity has widespread social ramifications. This is the political psychology face of an embedded situation.

People, social groups, and cultures resist changes that threaten a psychological dependency. Identification with an artifact that assumes symbolic proportions is often part of this. The reluctance of Maasai herders in East Africa to treat their cattle simply as commodities rather than as integral parts of the society is an example of this. The North American tendency to link personality and status to their automobiles also indicates some degree of psychological dependency on this particular mode of transportation.

Psychological dependency is important for two reasons. First, when people believe that something of value is being removed, a natural reaction is one of fear. Second, this fear is often exploited by political actors who wish to thwart reform efforts. Thus, it must be taken into account.

The example above of rural communities in the United States linking their identities to zip codes and athletic teams shows the power of this aspect of embeddedness. When ethnic groups feel threatened by social change they may evidence such attributes, as well. Any aspect of place, culture, beliefs, practices, symbols, or language that represent the core of a person's identity may indicate a psychological dependency. But these things are always in flux—no culture is static.

Fluidity

This brings us to the third aspect of embeddedness—*fluidity*. When societies experience political upheaval, revolution, natural crises, or drastic rearrangement of economic activity, one of two things may happen. Either people cling ever more desperately to the two aspects noted above, or they become loosened as people search for new sources of security. Intuitive political leadership may even influence which occurs. The key to successful policy reform is the ability to recognize opportunities emerging as part of a fluid situation. This is the temporal dimension of embeddedness.

Fluidity is an obviously pervasive factor in such places as Eastern Europe and the Newly Independent States today. Dramatic changes in economic and political systems challenge the embedded relationships that have evolved over the decades. But lesser disruptions can have similar repercussions for specific policies. For example, inundation with political refugees, the death or defeat of the nation's only post-independence ruler, the sudden collapse of the market for the major export, or other major events, economic shocks, or structural shifts can increase opportunities for implementing change. In Kenya, the burning of the elephant tusks and the enforcement of a total ban on the sale of ivory coincided with the elevation of the tourist industry to the number one earner of foreign exchange for the country. Leadership was poised to take advantage of an emerging resource dependency that favored protective action. Contextual change altered the probability that policy change could be introduced.

Embeddedness is important when assessing the changes that will be wrought by new policies. Even apparently simple things like the introduction of a different stove or fuel for domestic cooking may challenge embedded practices. For example, Marvin Harris notes that cow dung as a fuel provides exactly the right kind of flame for Indian cooking and any innovation that did not mimic it would have major consequences for the local cuisine and might encounter stiff resistance (Harris, 1974).

Other observers note various webs of socio-economic relationships in such diverse settings as East Africa and Southeast Asia. They suggest that the obligation and exchange relationships that permeate societies influence local reactions to what seem rational policies by outsiders. And they contend that these webs often promote behavior at odds with policy intentions (Scott, 1976; Popkin, 1979; Hyden, 1983). Embeddedness can distort policy impact. At the same time, practices and institutions embedded in the social customs and economic behavior of some ethnic minorities may help to explain their relative success (Landa, 1981; Jesudason, 1990). Thus building on positive embedded attributes may strengthen a policy implementation strategy.

With either type of embedded situation, fluidity is key. Whether a factor is on the increase or on the decrease may be nearly as important as its presence or absence. In Minnesota the fear of losing a zip code or a school is related to the trend for population growth to be centered in the Twin Cities region (actually a diagonal that goes from Rochester in the southeast to St. Cloud in the center and crosses through Minneapolis and St. Paul) while small settlements far removed from this axis are losing people.

To a policy reformer, then, the direction of a trend line for a dimension of context is important. The question is "is it becoming more or less embedded?" Will change become more problematic as the trend continues, or are there signs that opportunities for a shift in course are on the rise?

Likewise many natural phenomena follow cycles of abundance and scarcity, increase and decrease. Whether the last few years were drought years or wet ones can alter the perceptions, fears, and resistance to change by human populations. Political or economic events can affect migration, salience, power balance, or the strength of a resource decision system at any particular time. The difficulty of introducing policy change will be related to the point it enters an historical trend. Timing makes a difference. Thus, understanding embeddedness involves awareness of historical changes involving the resource base and the context surrounding it.

This chapter developed a map of context that can be used to clarify the setting that surrounds any environmental, conservation, or sustainable development policy. The map marks three major aspects of the setting. They are: (a) the problem context and its four dimensions of discreteness, progression, mobility, and boundary; (b) the social context and its six constituent parts of openness, power balance, salience, process, scale, and decision system; and (c) embeddedness with its three components of resource dependency, psychological dependency, and fluidity.

The route that a policy must negotiate through a setting, then, includes the fit between the policy and the nature of the problem, the match between the policy implementation strategy and the social dynamics that surround it, and the strength of the bonds among these items in terms of dependencies and fluidity. At any point an inappropriate fit can deflect the trajectory of the policy and lead to unintended behavioral impact. This pathway followed by policy through context is depicted in Figure 4–C.

We now have a context map. But two tasks remain before we can summarize the implications of this perspective. The first task is to test the context hypothesis by overlaying it on past experience to assess its explanatory abil-

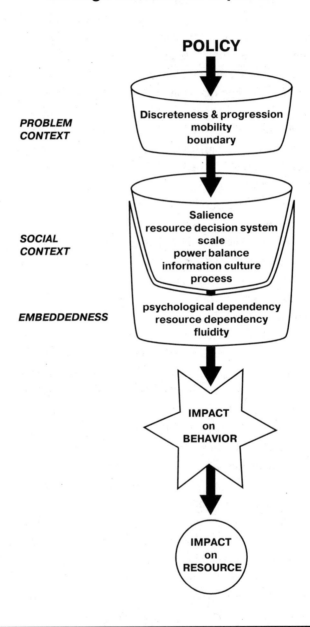

Figure 4–C
The Route a Policy Travels:
Through Contexts to Impacts

POLICY

PROBLEM
CONTEXT

Discreteness & progression
mobility
boundary

SOCIAL
CONTEXT

Salience
resource decision system
scale
power balance
information culture
process

EMBEDDEDNESS

psychological dependency
resource dependency
fluidity

IMPACT
on
BEHAVIOR

IMPACT
on
RESOURCE

ity. The second is to outline how to use it in the effort to craft processes for identifying and adopting future policies.

NOTES

1. Fred Riggs looked at countries undergoing a transition from what he called "traditional to modern" in terms of a ray of light passing through a prism. In the traditional form the light was white and the different dimensions of society (religion, politics, art, economy, etc.) were integrated, whereas in modern societies they were highly specialized and differentiated just as the light emerging from the prism was broken into the separate colors of the rainbow. But transitional societies he likened to the situation inside the prism where both the white light and the individual colors of the spectrum coexisted and flowed together. Based on this image, he developed an elaborate theory of "prismatic society." (Riggs, 1964)

2. In Liberia before the overthrow of the Americo-Liberian hegemony, for example, the civil service salary and wage payment system was designed to weaken morale, promote corruption, and keep upcountry people dependent on the largesse of the coastal elite. Parastatals, however, functioned as means for the elite to control specific resources and thus these "public" corporations worked under different rules.

3. An identical situation also exists for much state-owned land. For example, in Washington state the Commissioner of Lands must both preserve the natural beauty of state lands and manage them for income (personal communication from Brian Boyle, former Commissioner of Lands, State of Washington).

5

INTERPRETING THE PAST: TWO TESTS OF THE CONTEXT PERSPECTIVE

In the early 1980s I published an article on the topic of "rapid rural appraisal." That article was frequently cited in literature on that topic, but it was not cited for the reasons that I thought it would be. The article criticized many rapid appraisal approaches because they were not participatory and they did not try to enhance local knowledge. Instead, they were geared to informing outsiders. I thought this criticism was important, but it was seldom noted, let alone highlighted, when the article was referenced.

Also, over one-third of the text of the article was devoted to the contextual nature of the indicators that observers used and to the need to test whether or not they accurately depicted local circumstances. The argument was made that ". . . simple proxies can be misleading if their contextual fitness is not examined. . . . It is important . . . to articulate the assumptions that tie the proxies to the phenomena and to test these assumptions against local perceptions." (Honadle, 1982a: 636–37) But when the article was cited this point generally was ignored and the focus was usually on the indicators themselves—professionals were looking for the complete set of indicators to assess rural settings. Readers latched on to the indicators without questioning how well they traveled from setting to setting. Contextual considerations were not generally considered legitimate concerns at that time.

For contextual considerations to become legitimate, however, two conditions must be met. A contextual perspective should be able

- to show that it can help us to *reinterpret prior experience* and shed new light on reasons for success or failure; and

- to help us analyze specific circumstances and *devise improved strategies* for future policy reform.

In this chapter we address the first of these needs. First, we will use the context map as an overlay to examine an overview of some African experiences to see if it produces new insight about those experiences. Next, we will use the overlay to analyze, in more detail, a specific case of policy reform to see if it provides a more persuasive explanation for what happened than the common explanations. If the context map passes these tests, then we will have a reasonable confidence in its ability to help us to improve future designs for the implementation of policy reforms.

Revisionist history is sometimes an exercise in contextual analysis. For example, in the 1960s the management success of the Volvo plant in Malmo, Sweden, was attributed solely to the team approach introduced in that plant. But, at that time, the transfer of the approach to the United States was sluggish at best. Why was this the case? Context may contain the clue. Sweden was experiencing full employment and employers were in the position of vying for a limited supply of labor. In the United States, however, unemployment was higher and employers were not under the same pressure to compete for employees.

The late 1970s enthrallment with Japanese management underwent a similar context-based revision. A downturn in the Japanese economy led to layoffs in what was once considered a layoff-proof system. Cracks appeared in the seemingly impregnable wall of institutional loyalty and lifetime commitments. Although international organizations were enamored with the Japanese management approach in the early 1980s (World Bank, 1983), that rapture dissipated as the situation changed.

Other stabs at universal truth will probably experience similar deflationary reinterpretations. Although privatization, devolutionary management, and "reinventing government" solutions most likely contain some wise elements, time can be expected to show that they, too, fit better to some contexts than others. The question for this chapter, then, is "Does the contextual framework help to increase, or revise, our understanding of why things did or did not work?" To answer that question, we first turn to a review of local environmental action in Sub-Saharan Africa in the 1970s and '80s.

LOCAL ENVIRONMENTAL ACTION IN AFRICA

A book by Paul Harrison titled *The Greening of Africa* (1987) took the international development community by storm. Following close on the heels of a pessimistic assessment of Africa's environmental ills (Timberlake, 1985), Harrison's more optimistic study gave examples of the way that local action was helping to restore environmental quality in Africa.

The Harrison book provides an overview of numerous environmental action efforts throughout the continent. Some examples show how non-governmental, very local, initiatives led to change, as with the Kenya "green belt" movement. Others, such as the Niger forestry and land use planning effort, represent cooperative efforts among community, national, and international organizations. Although it contains successes, failures and mixed experiences, it is generally credited with an optimistic perspective on the potential of local action to reverse environmental deterioration in Africa at a time when most observers were pessimistic.

Even though no specific projects are treated in great detail, such a wide-ranging review provides a good starting point for a test of the utility of our framework. A perusal of this account provides a bird's eye view of many environmental initiatives in Africa with a stress on relatively successful ones. If, from this long range perspective, the context framework suggests some explanations for Harrison's findings, or indicates some avenues for further inquiry, then it has survived a preliminary test.

There is a danger in looking at "successes," of course—just because there is a pattern of characteristics associated with successes does not mean the cause has been discovered. Indeed, the same features may be present in failures, too. But at least this is a starting point for testing the value of a context perspective. We will identify some of the more prominent examples noted by Harrison and see if there is reason to believe that any of our contextual conditions offer interpretations for the level of success.

Although the focus of Harrison's study is on common attributes of many relatively successful efforts, there is a telling quote that shows that the link to context is important. According to Harrison, "Participation is . . . a way of making sure that projects are geared to local circumstances. In the African environment, culture and economy can vary from one village to the next more dramatically than in any other continent."(Harrison, 1987: 302) A question, then, is whether our social context categories help to understand the reasons for success.

Figure 5–A identifies examples from the Harrison book and explores pos-

Figure 5–A: Contextual Influences on African Local Action

Examples of Local Action	Contextual Influences
Kenya green belt	salience (+), scale/infra (+),information (+), RDS (+), power (+), process (+)
Zimbabwe maize	scale/infra (+), salience (+), RDS (+)
Niger Majjia valley	power (+), RDS (–), salience (+), process (–)
Niger forestry and land use planning	process (–), salience (+), power (+)
Ethiopia regional integrated basic services	power (–), salience (+), scale/infra (+)
Burkina Faso *naam* movement	RDS (+), salience (+), process (+)

sible contextual influences in terms of support for strong outcomes (+) or explanations for why partial failure occurred (–). Only social context categories related to discussion in the book or other knowledge of the locale and example are included. The treatment of the projects in the book did not provide information needed to overlay the problem context dimensions, and thus the discussion is limited to social context categories. However, where salience was high, thresholds may have been crossed and problem context dimensions may be important without being apparent.

The six examples were chosen due to their prominence in the book and they were not in any way screened to reflect inclusion of information about context dimensions. Even so, the context map is useful—six of the most prominent cases in the book benefit from overlaying the social context template. Each of the context components is noted below with the number of cases where it illuminates reasons for positive or negative results.

Component	Cases (of 6)
Information	1
Power Balance	4
Salience	6
Process	4
Scale/Infrastructure	3
Resource Decision Dystem	4

This application of the context map suggests that it does help to raise some interesting questions about what was responsible for the success levels of these examples. Dimensions with positive influence seem to be important in explaining why they worked as well as they did. And when Harrison questions either the immediate impact or the longer term sustainability of specific efforts, the questions focus directly on key contextual dimensions that are captured in our scheme.

Since Harrison's book was not written with these dimensions in mind, the data base is spotty for our purpose. Sometimes I had to resort to my own knowledge of the country where the project was. Even then I could not develop a coherent view of the problem context. The social context dimensions were easier to identify. Each element of social context is represented in some of the six cases, and one element appears in all of them, suggesting that the context model should not be rejected without cause. The most that we can conclude from this exercise is that the context template holds promise. We cannot conclude definitely that it works, but we can infer that it is worth pursuing further.

Having passed this initial rough assessment, a higher hurdle is appropriate. The question is, "Can the context framework provide insights to explain any of the details of, or offer an alternative explanation for, a more extensive case study?" To answer that question we turn to Asia and to a mini-case study of a reform effort that I have some personal field knowledge about—the National Irrigation Administration in the Philippines.

THE PHILIPPINE NATIONAL IRRIGATION ADMINISTRATION

Water management is a key problem area in natural resource management and sustainable development. Indeed, the first annual conference on "Environmentally Sustainable Development" held at the World Bank in 1993 treated the water management issue as central (Mulk, 1993). Fortunately, one of the most well-known cases in the development management literature involves an irrigation management agency, and the experience with this agency lends itself to a contextual re-examination.

The reorientation of the Philippine National Irrigation Administration (NIA) from a top-down builder of irrigation infrastructure to an organization experimenting with participatory irrigation management is one of the success stories of international development administration. The leadership of NIA senior administrator Benjamin Bagadion and other Philippine colleagues was critical to the NIA's adopting a participatory approach to its irrigation development. Technical assistance from Frances Korten of the Ford Foundation and the writings of David Korten added support and inspiration for this effort and even turned an international spotlight on it.

The agency's transformation was remarkable. The NIA was established as a semi-autonomous public corporation "to investigate and study all available and possible water resources in the Philippines, primarily for irrigation purposes; to plan, design, construct, and/or improve all types of irrigation projects and appurtenant structures; to operate, maintain and administer all national irrigation systems" (National Media Production Center, 1977: 75). NIA began in 1964 with a technical orientation and a culture to support that orientation. Its role was to bring the Philippines into the "irrigation age" by building and managing large command areas, charging the cost of operation to the farmers who benefitted from the systems, and recouping the costs of construction.

But by the mid-1970s NIA was beginning to reconsider its role. Presidential Decree No. 552, issued in September 1974, allowed NIA to keep revenues collected from irrigation fees, rather than returning them to the general treasury accounts, and it empowered the organization to delegate the management of its irrigation systems to local associations. Agency leadership realized that this could be a much more cost-effective way to manage the systems. But that would be the case only if the associations' capacities were sufficient to meet the challenges of system management.

Consequently, NIA, with Ford Foundation assistance, embarked on a program to build its own capacity to promote the emergence of able local associations to manage the physical infrastructure. But this was a daunting task. NIA was used to building the physical systems using technical criteria, not to enabling the development of local groups or to synchronizing social and physical infrastructure development.

To bring about this change, working groups were established at various levels within NIA and process documentation and community organization methods were introduced at the field level. After an intensive reorientation effort, NIA was transformed into an organization with a new set of tools to complement its civil engineering world view. In the pilot areas the relationship between NIA and villagers was altered dramatically.[1] This experience was examined in detail and the picture that emerged from the examination became the accepted wisdom about what happened, why it happened and how to do it elsewhere.

The Conventional Interpretation

The accepted view of why the NIA reform succeeded is that three elements—the policy framework, the methods used, and the management process were appropriate (see F. Korten, 1988). In terms of policy, the legal status of irrigator associations and the ability of NIA to recover system operation costs from those associations set the stage for innovation (see Bagadion, 1988). But legal and financial policies were noted as important only in retrospect. At the time of the reform the innovation process itself was the focus of attention. There were two major reasons for this: (a) because the legal and financial policies were in place by 1974, before the major thrust of the reform program began, and (b) because the vantage point of the facilitators and observers stressed internal organizational characteristics and the reorientation process.

Indeed, organizational factors were important—factors such as NIA's relative autonomy combined with the congruence of the command area of the irrigation schemes and the membership of the associations. These helped NIA to impose a new internal operating style and helped the irrigator groups to build their capacities. But this too was background. The real attention went to the methods and the management process.

David Korten's 1980 article describing a three-phase "Learning Process Approach" provided a clarity that previously had been lacking in the field of international development (Korten, 1980). He found that a sequence of first learning how to do something, then learning how to do it well, and only then expanding to do other things characterized successful efforts at building local organizational capacity. This became the core of the organizational development technology employed in the NIA transformation. And it also became the basis for interpreting what happened.

Various mechanisms and instruments were developed to support the installation of a learning process approach in both the farmer associations and the NIA itself. Participatory processes were introduced by using working groups to involve actors and guide the change process, by using socio-technical surveys to map the interface between community dynamics and system needs, and by using process documenters to record the process and provide a basis for discussion and organizational learning.

In fact, this was a very management intensive exercise. And it was a new experience for engineers—recasting villagers as colleagues rather than as underlings who would use the creation of the professionals. Meetings, workshops, and collaborative exercises characterized work at both the agency and the community levels. And the international professional attention given to the NIA experience elevated it to the status of a global model for process approaches.

A process approach was a step forward from previous approaches to development administration that posed a comprehensively blueprinted design as the key to success. A process focus was dynamic and evolutionary as opposed to the static blueprint model (Sweet and Weisel, 1979). This new model substituted participation for imposition and an evolving process for a predetermined design. The story of the NIA case is the story of how this was done. The attention was on the intervention itself as the determinant of impact. The assumption was that if it is done right it will work here, and if it works here it will work elsewhere. The process was the key to success.

But, as is so often the case, the interpretations of the reasons for the success are guided by the observation point of the observer. Those who study process tend to see process as the driving force. Those who do process work tend to recommend process as the answer to difficult circumstances. This is not meant to belittle process. Indeed, this writer has emphasized the need for appropriate process in many circumstances. But if a contextual observation point were used, the interpretation of the reasons for success might shift away from the reform activity itself toward an improved understanding of the receptivity of the setting to the initiative. This, in turn, would highlight the fit between the two.

The Context Revision

Our context map contains some landmarks that help to interpret the reasons for the NIA success, while others offer less assistance. Some of the problem context categories, for example, are geared toward a situation where a natural resource is in a condition of decline, and the objective of the policy is to improve that condition. In this case, however, the situation is different—neither water quality nor quantity is at risk. Thus, the categories of discreteness, progression, and mobility have limited application to this experience. But the boundary category does apply.

The fact that the irrigation association boundaries matched the command area boundaries greatly facilitated the process. In fact, an attempt to replicate the NIA process in Indonesia was complicated by the fact that the management organization units followed village or settlement boundaries and not the boundaries of the water systems (personal communication from Frances Korten). So one of the problem context dimensions focuses our attention on a fit factor and offers a partial explanation for the NIA success. But social context categories might improve our understanding of why things worked as well as they did.

Scale, space and infrastructure: Surely a decentralized approach has advantages in an island nation. But NIA's hierarchical engineering style often resulted in field projects bypassing regional offices and reporting directly to a central projects office (and even keeping in constant contact via short-wave radio). The road infrastructure on Luzon was good, partly due to roads built by the Japanese as part of the World War II reparations. But, with one important exception, this dimension does not seem to add significantly to an improved explanation of the NIA experience.

That exception relates to organizational density and capacity as an element of infrastructure. The number of university graduates, the number and distribution of NGOs specializing in community development activity, and the quality of many of the programs already operating made the Philippines a resource rich location. When NIA needed to hire new staff with new skills, there was a pool of them to tap. And they were already active throughout the country.

Thus, this social context dimensions adds something important to our understanding. And the other five may be even more useful for new insights into why the NIA experiment worked. Likewise, the concept of *embeddedness* may help to clarify the reasons for success.[2]

Resource decision system: The *resource decision system* role is more complicated than the simple boundary perspective. The match of irrigation association boundaries to the command areas of the water systems certainly made capacity building easier because building associations along physical boundaries escaped from some of the entangling dynamics of village authority structures. Likewise, NIA's structure as a public corporation gave it some autonomy to experiment with new approaches and its location outside the civil service made the introduction of reform more manageable than might have been possible under other circumstances. Moreover, the policy change that allowed NIA to keep the revenue from the local water management associations provided a direct incentive for the agency to make the associations viable as management organizations.

Multiple elements of the decision system thus reinforced the reform process. The fit between the structure of decision-making within NIA and the intervention, then, probably is a partial explanation for the level of performance. In fact, the use of work groups and engagement of key people can be presented as a direct attempt to match the intervention process with the resource decision system. The process was good.

But this offers only a slight reinterpretation of the traditional explanation for why it worked as well as it did. The change agents were sensitive to the pressure points within the resource decision system and the method was meshed with those points. What this overlay does provide, however, is a partial explanation for why a process approach works—a participatory and evolutionary learning procedure can explore and develop ways to fit a reform to a resource decision system. Indeed, it can even facilitate the mutual adjustment of the decision system and the reform effort to each other.

Informational openness: The Philippines is a relatively open culture. A wide range of printed political views were available even during martial law. And many Filipinos are free-wheeling marketeers, scorning hierarchy and formal channels. At the same time, the Spanish influence with a more rigid, Roman Catholic, status-based view of the world is strong. Prefixes indicating doctor (Dr.), or attorney (Atty.), or engineer (Eng.) are regularly used in the news media to assign professional and class status. Indeed, this cultural duality appears as a theme in Philippine fiction (Joaquin, 1972).

But, on balance, the informational openness of the Philippines provided a background supportive of alternative approaches. The existence of more than one party (even though elections were frozen during the period of martial law), the availability of social criticism (such as that at *Solidaridad* bookshop in Manila), the willingness of subordinates to give critical feedback to their superiors, and open discussion of political issues all facilitated the promotion of participatory methods.

And the reorientation of NIA was information-intensive. Process documentation and the involvement of local association members in agency processes required vast amounts of information sharing. This would have been more problematic in more inscrutable societies. In fact, the country had a large community of non-governmental, private and non-profit organizations involved in community development activities. Sharing information was central to their *modus operandi*. Indeed, the processes used by community development agencies resonated with those introduced into NIA operations. Thus, the informational dimension of context did not inhibit the performance of the reorientation effort and it affected it quite positively.

Process requirements: The organizational change technology used to redirect NIA was derived from the organization and management sub-field of organizational development (OD), or applied behavioral psychology. Among the major features of OD-type approaches, three are important for our purposes. They are:

- people's *perceptions* are considered legitimate data that are vital for management decision-making;

- workshops, task groups and *interactive, participatory exercises* are used to identify diverging and converging objectives, build coalitions, and choose management strategies; and

- *facilitators and process documenters* are used to ensure the momentum of the process and to enhance conscious awareness of it.

Although OD-based analytical approaches have worked in a wide variety of settings because they build on local perceptions (Cooper, 1987; Mbise, 1987; Silverman, 1987; Honadle and Cooper, 1989), nevertheless, some cultures receive them more enthusiastically than others. The Philippines is one of the most receptive places in the third world. Indeed, OD activities were so common before the NIA exercise that a "T-shirt phenomenon" was recognizable. That is, OD workshops were marked by the printing and distribution to participants of T-shirts adorned with the title, date, sponsor and place of the workshop. Thus, the general approach was particularly suited to the process requirements of Philippine culture. But this alone cannot explain the success.

Power balance: Another important social context factor was power balance. NIA was a very strong organization capable of overcoming the opposition of organizational competitors for resources or clients.

A trip to the NIA headquarters in Quezon City left a lasting impression in the mind of any visitor—this was a high-capacity organization compared to other Philippine government organizations. The staff consisted of highly trained people (many engineers), the physical facilities (headquarters, regional, and project offices) were superb compared with the shabby offices held by such departments as Agrarian Reform, Plant Industry, or Community Development, the communication equipment worked and was independent of the national phone system (Short Wave radio made it unnecessary to rely on the phone system), high-quality transportation equipment was available (4WD vehicles, lots of vehicles per staff member, vehicle service facilities directly under NIA), and staff were relatively well paid (high budget per staff, relatively low percentage of budget consumed by wage bill, money available to hire social scientists at competitive rates). These indicators all supported an assessment that NIA was a strong organization compared with its competitors.

Indeed, at the field level this was very obvious. Comparing the resources available for different Integrated Rural Development Projects (IRDPs) in the

Bicol area of Central Luzon gave a stark picture. The NIA IRDP at Libmanan-Cabusao eclipsed by far the settings run by other agencies, such as the Ministry of Agrarian Reform project at Bula-Minalabac. It was clear where the human, organizational and physical resources were concentrated. And such a concentration of resources raised the chances for success.

But more was needed. Without an organizational commitment to depart from previous approaches, success still was not likely. This was provided by a conflux of individual and organizational imperatives.

Salience: Salience was high both in terms of individual receptivity to innovation and in terms of organizational awareness of the need for new thrusts. The individual was a key actor, and senior NIA official, who was involved in promoting and leading the reform initiative. He was receptive to a new approach due to a number of converging reasons. One of these was recent experience with NIA in his home town (Libmanan). This experience was seen as negative by locals—NIA was depicted as a strong-arm actor espousing participation but pushing through its own agenda. Its walled-in compound was seen as excluding villagers, and the style of the project manager was seen as authoritarian. The NIA official had a direct link to the community. In fact, his brother, an attorney, served as acting mayor during a trip overseas by the mayor.

And his willingness to promote change had other foundations, too. For example, the requirement that farmers pay back to NIA the cost of improvements made participation necessary because farmers resisted paying for improvements that they did not want. This made it necessary to involve the system users in decisions about system refurbishing and system design. He understood this implication. Additionally, his son was involved in community organization activities and had sparked his father's interest in the potential of this approach. Thus, a key individual saw the need for agency reorientation.

The organizational awareness related to another issue—NIA had jurisdiction over large-scale national irrigation system development only. But the country was running out of places where large-scale systems made sense and so NIA looked at the smaller communal systems as places for future work. But the communal systems, which were generally under two hundred hectares, were assisted by the Farming Systems Development Corporation (FSDC) which had a community-based approach to system design and management. FSDC engaged in capacity-building with the local organizations which actually ran the systems. NIA, on the other hand, built and ran the national systems with local people paying for the service. But if NIA were to develop associations to run its command area systems, then it would need expertise in building the capacities of those organizations. FSDC had that expertise. NIA did not. And successfully assuming this new mission would require a different operating style.

Also, irrigation was for rice production. For political and economic reasons the Philippines wanted to achieve self-sufficiency in rice. Moreover, the Philippines exported rice even though some areas of the country were food deficit areas because some of the elite sold rice to earn foreign currency in the lucrative Asian rice market. Thus, both the drive for self-sufficiency in rice production and the desire of some of the Marcos-supported elite to use rice to earn foreign exchange kept NIA as a favored agency.

The participatory approach gave NIA what it needed—a way to enter a new market for its skills and a justification for going back and retrofitting its old systems with a new process technology. Now it had a new lease on its mission.

The social context dimensions help to explain why the reorientation effort worked. But there is also the question of embeddedness.

Embeddedness: NIA's *resource dependence* was not threatened by the reform effort. If anything, the reform acted to reinforce the organization's sources of support in the professional community, the upper classes, and the government and it also improved its image among the village masses. In fact, international attention to the effort even made it a magnet for international funding agencies who wished to claim some of the credit. Additionally, by going into partnership with local irrigation management associations, NIA expected to strengthen its financial position. Thus, the reform occurred in a situation where resource dependency was not a stumbling block.

Psychological dependence is a more complex issue. Elevating irrigator needs, skills and roles to a more prominent position certainly challenged the engineering view of the world that dominated NIA. Thus, there was a threat to self-image in terms of the all-knowing expert and socially dominant professional. But, at the same time, the presentation of new skills as cutting-edge technologies and the promotion of these skills within NIA served to cast it in the same old light as a premier organization. Indeed, as foreign visitors came to view NIA's new participatory initiative, organizational pride took the place of resistance. The use of task forces also served to involve the power structure in the transition to a new operating style. Thus, the change strategy assumed a degree of psychological embeddedness and worked to overcome it.

As the discussion of salience noted, the situation was *fluid*. Changes in the organizational environment supported a loosening of embeddedness and facilitated the change effort. Two component parts of embeddedness, then, supported change, and the third was directly engaged in the reorientation exercise.

Thus, the various contextual dimensions had either positive or neutral impact on the reform effort—none evidenced negative effects (see Figure 5–B). The fit between context and intervention—not just characteristics of the intervention—provides a more convincing explanation for performance than traditional assessments of the attributes of the implementation process alone.

The NIA case documents the superiority of process approaches over blueprint approaches for building self-management capacities in local organizations. That has been its historical significance. But it also provides information that allows us to move away from a focus on intervention style alone to focus on the match between the intervention and the context. It contains lessons relevant both to the move from blueprint to learning process and to the move from implementation process to contextual fit. Indeed, the context perspective is not a return to a blueprint mode, but rather an advanced step toward the process of fitting policies and programs to their settings.

Retrospective examinations by the people involved in the NIA experience support the context perspective. When the reformers were enmeshed in the change process, they credited that process with the results that were achieved. But, after moving on to new places or new activities, they began to see more clearly the important role that context had played in determining the outcome of the reform effort (see, for example, F. Korten, 1986).

This reinterpretation of the NIA experience supports a contextual view— it suggests that the fit between a context and a reform process provides a more accurate interpretation of reasons for success or failure than a focus on the reform alone. Thus, the context map has passed a second test—it has exhibited explanatory power when applied to this case study.

The context template has now passed two tests: first, it provided an explanation that is consistent with the findings of a cross sectional review of African experience; and second, it provided an improved explanation for what happened in a more detailed Asian experience. Thus, it has face validity as a reasonable interpretation of why implementation failed or succeeded. Furthermore, the context map explains, in a systematic way, why similar interventions might fail in different settings and it reveals why a simple emphasis on the replication of interventions is a strategic error—the context itself must be in the spotlight because it is the fit between context and strategy that produces outcomes.

Explanation is important. If we do not know why things happen it is

FIGURE 5–B: Contextual Influence: The NIA Experience with Bureaucratic Reorientation

Contextual Dimension	Direction of Influence	Nature of Influence
resource-threat relationship discreteness progression	neutral or not applicable	although water flowed, the management systems were site-bound and contained the water; effective water use could be facilitated by local self management
boundary congruence	positive	NIA had national authority; in some irrigation systems multiple municipalities were involved but water provided the system boundary.
information/political culture	positive	the Philippine culture supported information sharing
process values	positive	organizational development approaches had already been embraced and incorporated into local management practices and this supported the style of the intervention
decision system	neutral or positive	use of working groups accommodated to the power structure of NIA; legal authority for irrigation associations was appropriate; association boundaries were appropriate
salience	positive	both key individuals and the organization saw this as a salient opportunity
scale/infrastructure	neutral or positive	scale did not appear to influence implementation but organizational density and capacity were supportive
interorganizational balance	positive	NIA was a very high capacity organization compared to any potential competition or opposition
embeddedness - resource	positive	no threat; reinforced support
- psychological	neutral	threat offset by cutting edge perspective
- fluidity	positive	changing mission supported the innovation

much more difficult to make them happen again. But explaining the cause is just the first step. The next step is to take our knowledge and turn it into a method for inducing the results that we want. Since the context map stresses the importance of certain categories of local circumstances, what we now need is a format for applying that knowledge to new settings. We need to be able to use the context map to design policy reforms, to fit those reforms into a change process, and to make strategic choices that result in desired outcomes.

NOTES

1. Even though some observers contend that the transformation of NIA was incomplete and geographically limited (Ostrom, Schroeder and Wynne, 1993), nevertheless, this experience should not be dismissed. It was deserving of its reputation as a success. This author has examined numerous "successes" and found fault with all of them (Honadle, 1986). All success stories have incomplete or failed elements, just as most "failures" contain successful components or positive outcomes (see Hirschman, 1967).

    However, there may be another lesson here. When the national policy setting contains contradictory policies residing in different clusters, the staying power of a successful implementation may be short. High salience may support an initial innovation, but once the initial effort has been made, policies supporting routine and adaptive behavior may need to come into play as salience declines. Indirect incentives that exhibit low insulation may be especially important. Although exact sequences may vary from situation to situation, if the end result is not a supportive policy environment then success stories may become little more than quickly disappearing blips on the screen of history. For excellent examinations of a wider range of natural resource issues in the Philippines, see: Broad with Cavenagh, 1993; Dauvergne, 1997.

2. This assessment is based on the author's field experience in the Philippines on multiple visits during the period 1977–81. (See, for example, Honadle, 1978)

6

TOWARD DESIGNER POLICY PACKAGES: CONSIDERATIONS WHEN USING THE CONTEXT MAP

In the mid-1970s I served as a member of a team designing an Integrated Rural Development project in West Africa. The team consisted of consultants to and direct-hire staff of two international donor agencies.

My role was to determine the land holding system in the project area, to propose the organizational structure that would implement the project, and to provide an outline of management mechanisms and processes that would be used. To do this, I needed to know how the local institutions worked and I needed detailed information about bureaucratic dynamics and procedures, land tenure, social organization, and political stumbling blocks.

Because of my training I was aware of various organizational options and combinations thereof, and I knew that I needed to listen carefully to what local people told me or I would misinterpret what was happening and any proposals I made would miss the mark and project performance would suffer. So I listened to people's views of what would and would not work and their explanations for why each was so. But when I related my conversations to the team leader at an evening meeting, I received a surprise.

I was informed that the basic organizational structure was known—the major donor always used Project Management Units for rural development projects in this part of the world. It was the preferred organizational form, based on a civil engineering model, and all I had to do was develop the details and the implementation strategy. My protests were to no avail. Organizational preference for a certain style of operating outweighed situational differences. All I could do was elaborate variations on a theme and a succession of organizational configurations.

To be fair to the team leader, though, the problem was not just a matter of an organization's preferred approach. It was also a result of a lack of tools to analyze alternatives in light of contextual conditions. And that lack remains to this day.

In the 1990s I was on a forestry sector mission in Asia where the team leader from the funding agency was using a policy analysis matrix (PAM) to try and identify policy issues that could affect the implementation of a sector loan. But this tool was markedly ineffective and it was mute testimony to the need for improved frameworks for identifying policy implementation options and relating them to local circumstances. The PAM was a blunt instrument in a situation that called for a scalpel.

There is a continuing need to construct assessment tools. Given the preliminary performance of the context map, perhaps it could be used to build an analytical framework or to outline a process for fitting policy to circumstance.

Indeed, if environmental policy formulation and implementation are so important, then it behooves us to do them well. If context is a major determinant of implementation success, then addressing context issues will be an important part of the process of designing an implementation strategy that will work. A key element of any such process will be identifying what aspects of the present setting can be built upon and what ones need to be replaced altogether.

REFORM VS. REDESIGN

The difficulty encountered when introducing policy changes relates to the embeddedness found in all context dimensions. And it also relates to the nature of the change—does it go against the flow, or does it go with it; does it just reform or modify what is already there or does it eliminate something and substitute something totally new? Another way of describing this interaction between the degree of innovation and the context is to draw a distinction between what has been called "trait-taking" and what has been called "trait-making." (Hirschman, 1967)

Trait-taking

A trait-taking reform has at its core an effort to build on traits that are already present in a setting and it strengthens or uses them. A trait-making initiative, on the other hand, primarily attempts to instill entirely new behaviors or characteristics or substitute new ones for old ones. This can run counter to ingrained practices, interests or values. Needless to say, trait-making is more costly, difficult and problematic and it will be more likely to work when embeddedness is loosening.

A hypothetical example can help to illustrate the difference between making and taking traits. A regulated utility company is ordered to engage more resources in electricity demand management rather than assuming that increased generation is the answer to projected shortfalls in the near future. One way to do this is to accelerate the spread of flourescent light bulbs to replace incandescent ones. The company will underwrite the cost of the transition to the new bulbs by making them available at a highly subsidized rate. It has been determined that this is very cost-effective compared with increasing generating capacity and it can more than compensate for the expected shortfall. Two options are to be considered for doing this.

Option A is for the utility to provide the bulbs to customers at a subsidized rate. This would require a new unit within the organization to acquire and distribute the bulbs. Information could be channeled to customers along with monthly power bills. Although it uses an organization that already exists, this approach is trait-making because it causes an organization set up for one purpose (the provision of electric power) to perform another unrelated one—retailing a hardware product.

Option B is for the utility to provide discount vouchers that can be redeemed at local hardware stores. The vouchers would go with the monthly bills, the local hardware stores and their distribution networks of wholesalers and retailers would perform their normal functions, and the power company would contract to a specialty firm to handle the payment to the hardware retailers.

The Option B alternative is more trait-taking than Option A because it uses channels already in place to perform the new functions. Each organizational role is occupied by an organization that already performs the respective function. This option would also encounter far less resistance in the local context because it benefits pre-existing organizations rather than placing the utility in (unfair?) competition with them.

Opportunities for trait-taking are chances to improve the effectiveness of an implementation process. The ability to identify trait-taking opportunities is one that can be sharpened. If trait-taking makes implementation less risky and more easy, then it would be useful to have some guidance for identifying traits that offer potential as reform channels. What kinds of things might be augmentable traits? The following offer some categories of traits to be considered:

- *Values*—are there symbols, processes, roles, psychological factors, or cultural practices that generate loyalty or respect and that could be employed in the change process?

- *Functions*—are there education, communication, production, distribution, disposal, maintenance, governance, religious, social, recreational, employment, research, or other functions that are presently being performed by people or institutions that could perform an allied, parallel or identical function during the reform effort?

- *Structures*—are particular organizations in place to reach subsets of the population, is it necessary to follow particular channels when conducting certain types of business, are specific pre-existing institutions the only ones that can legitimately do certain things, and are any of them crucial to the implementation strategy?

- *People*—are particular leaders able to mobilize support for new departures, do some groups have more self-interest in promoting, accepting, or engaging in the reform than others, are there non-traditional power centers that are emerging as a result of salience or the untangling of a previously embedded situation, and could any of them either spearhead or support the policy change?

Answers to these questions might redefine what appears to be a situation requiring trait-making to one where trait-taking could be employed. And this reorientation could lower the level of conflict and improve chances for successful performance.

Trait-making

There will be times when radical change is needed and trait-making is essential. Systemic disruption may be an objective of the new policy initiative. Sometimes the problem will not be solved without a complete turnabout—not just a slight deviation from course. It is nice when the changes that a policy induces are peripheral to basic values and ways of doing business in a society, or when the innovation can build on core practices, because this makes implementation easier. But the more the impact of policy reform involves changing strongly embedded behavior, or behavior that is becoming more embedded, the more difficult implementation will be and the more important the inclusion of key stakeholders becomes.

Many things may force a trait-making approach. For example, the need for radical redesign may be a function of salience due to crossing a threshold—the old practices may not be viable in the new situation. When either local resource management or incentives failed to work, the need for command and control may be obvious. Conversely, the inability of command and control to make a difference because of the diffuseness of a threat, the inadequacy of resources, or some other factor, may lead to an array of strong direct and indirect incentives to accomplish the task.

Thus, trait-taking or trait-making strategies will not conform to any choice of policy type. Rather they relate to the degree of substitution needed to change behavior, and the systemic consequences of the behavior change. Interdependence of behaviors may be so great that only a radical reconstruction of relationships will save the day—systemic change may require trait-making.

When radical change appears to be the only viable approach, some people will welcome that reconstruction, but others will mobilize to defeat it. The level of conflict will escalate. In such circumstances interorganizational power balances, capacities related to scale and infrastructure, and the configuration of the resource decision systems are likely to be important factors affecting a change strategy. Indeed, the relative capacities of different stakeholders will shape the implementation terrain.

STAKEHOLDERS

Overcoming an embedded situation requires the injection of new incentives, institutions or practices that overpower present ones and substitute the desired ones for the undesired ones. This is seldom an easy task. People with a vested interest in perpetuating the old system will mobilize to thwart the reform effort. Others, who see the change as positive, will support the reforms. Still others will watch the drama and be either amused or dismayed, enlightened or antagonized, inspired or angered. All of these people may be seen as stakeholders.

Satellite Stakeholders

Different actors can be stakeholders for different reasons. A stake in an issue may arise from a fear of losing wealth, power, prestige, authority, or convenience. Or the stake may be based on professional or amateur interest or personal concern about potential change. In some cases, chances for personal or institutional gain may be behind the interest. And, as noted above, psychological or resource dependency may be what drives the concern.

The array of potential stakeholders for any policy or reform, then, is large. Not only does it include actors in the immediate vicinity of the proposed action, but it may also include organizations and individuals located far away. For example: furniture makers in Europe may have a financial stake in policies that affect the supply of tropical hardwood from Malaysia; residents of Chicago may be concerned about oil drilling in Alaska and its affect on caribou migration simply because they appreciate the beauty of the area and the value of vast expanses of unspoiled tundra and the caribou symbolizes this to them; fishermen at the mouth of the Zambezi River in the Mozambique Channel may be concerned about water management practices in Lake Kariba eight hundred miles upstream; or the New York Zoological Society may have a professional stake in policies that affect rhinoceros protection in Kenya. Indeed, anyone who perceives that they have a stake in the situation is a potential stakeholder. They do not need to be directly involved in the management of the change to hold this perception.

Such actors and observers can be called *satellite stakeholders*. Two things characterize them: first, they may be more spectators than players in reform efforts; second, they are an outer ring of interested or informed bodies that may support or oppose direct actors.

Any initial assessment of the universe of stakeholders will include some that are satellites. Covering that universe involves identifying those who possess the different kinds of stakes in any outcome. Nine categories of stakes need to be considered as potential sources of actors that can influence the course and results of implementation strategies. They are:

- *intellectual*—some people are committed by belief, interest, personal identification and reputation, or other factor to a particular definition of a problem, to the solution of a specific problem, or to certain solutions to problems. Thus they have an intellectual stake in some issues;

- *appreciative*—some actors have an interest in an issue because they simply appreciate a resource. Many may not use it but they appreciate its being there, others see the potential to generate information and a chance for learning, others may like a view or an animal for non-consumptive purposes. All of these stakes are grounded in appreciation;

- *organizational mission*—some organizations may have a charter or history that gives them stakes in a problem area (for example, a transportation authority, a community development society, a pollution abatement agency, an agricultural technology research center, a land use commission, a zoning board, a tax reform institute) and they will want to be involved in any discussion of changes in policies affecting the area;

- *user*—the users or consumers of a good or service that could be changed in nature, source, availability, price, or timing by a policy shift will be concerned about any policy change. They have a direct stake in any outcome that produces change or perpetuates the former situation;

- *provider*—those who provide or market any good or service affected by a policy may gain or lose from any change. Thus they are direct stakeholders;

- *power*—some actors may lose sources of power or influence over others or over their own lives. The threat of lowered autonomy, as in the case of landowners or the proprietors of businesses, or a chance to increase control over other actors can create a stake in any situation;

- *status*—a policy or its means of implementation may change the roles played by some people, and this can change their social, political, or professional status relative to others, and in some cases it may affect their sense of identity. Those who gain status and those who lose status may consider themselves stakeholders in alternative outcomes;

- *burden shifting*—many present policies allow people and organizations to transfer some of the cost of their operations to others. Those who are the passers of costs and those who are the receivers of burdens are all stakeholders who stand to gain or lose from change; and

- *legal*—people or organizations may gain or lose legal rights or protections as a result of changes in a law, elimination of a law, or introduction of a new law. These actors have a stake in the outcome of any reform process that would change their legal status.

Policy reformers need to know the approximate and relative numbers, intensity of interest, and capacities of actual organizations and individuals that represent each category of stakeholder interest for any specific initiative. Once this has been determined it is then possible to derive a list of key stakeholders for inclusion in the reform process.

Nuclear Stakeholders

An age-old question that has plagued managers is "How do you get everyone in on the act and still get some action?" In today's international global economy there is another potential stakeholder around every bend in the road, there is a general consensus that it is good to use participatory processes for ethical reasons, and, at the same time, the immediacy of many resource threats propel us to act more quickly than in the past. Thus, the question is perhaps even more germane today.

The stakeholder universe must be pared down to a manageable set. It cannot be too small without running a risk that people with key knowledge or needed commitment will be omitted. But if it is too large then the process will become unwieldy, momentum will dissipate and little will result. There must be a way to glean the central people and organizations—*the nuclear stakeholders*—from the array of satellite stakeholders.

Some categories of satellite stakeholders may contain nuclear ones. But some may not, depending on the specific reform and its context. A simple criterion can be used to find the nuclear stakeholders in each category. They will be:

- those actors who control resources needed to implement the change; and

- those actors who can mobilize resources to block it.

Overlaying this perspective on the master list will produce a shorter list of those whose involvement will be important for ensuring success. The overlay should be done by a group of people informed about the setting and the problem. Analytical exercises can be used to identify where the participation of each nuclear stakeholder is crucial (Silverman, 1987; Honadle and Cooper, 1989).

Developing a stakeholder profile makes it more possible to overcome embedded circumstances. Stakeholders are central to the development of an implementation strategy and they are integral to the social and problem contexts that surround any effort at policy change. Analyzing each dimension of context produces insights into who they are and what roles they might fill.

RETRACING OUR STEPS—A PROCESS

Looking back at the progression of the discussion to this point may help to see how the context map suggests a policy formulation process. In the beginning is a problem—one that is already causing difficulties, or is about to cause difficulties, or is anticipated to be serious in the future. Commonly that problem involves some type of decline in the condition of a natural resource. The *problem context* elaborates upon the nature of the problem. That nature includes how independent of other phenomena the threatened resources are, how the severity of the problem is progressing, the nature of the interaction between the resources and the threats to them, and the arrangement of social boundaries that surrounds the interaction between them.

Given this problem context, we can identify some set of human behaviors that need to be stopped, changed or introduced to lower the threat and improve the situation. These *behavioral objectives* for policy fall into two major categories—either they are specific actions by specific actors that can be targeted for reform, or else they are much more amorphous effects of systemic interactions by the combined behavior of many actors.

Once we know the key behaviors then we need to see what is causing them and what may stand in the way of changing them. This can be elaborated by exploring the component parts of what we have called *social context* and *embeddedness*. The social context categories of political culture, process requirements, power balance, salience, scale and infrastructure, and decision system all expose elements of the situation that need to be understood before embarking on policy reform. If the policy is based on a misinterpretation of these elements, then it increases the probability that it will fail. And the resource dependency and psychological dependency of the people whose behavior will be affected also are very important considerations when formulating a policy strategy. Indeed, these categories reveal potential implementation barriers for each of the units within the boundaries made explicit in the problem context. And, of course, the fluidity of the circumstances is important for understanding the difficulty and resource requirements for introducing change.

With this description of the setting before us, we can examine the four broad categories of *policy options* to see which option, or what combination of options, best meets the needs of the situation at hand. Command and control, local resource management, direct incentives or indirect incentives may all have roles to play. The question then becomes one of selecting the most appropriate *mechanisms to implement* the policies and matching the means to the objectives. But this selection process is circular and simultaneous—not just linear and sequential. That is, the choice of optional policies and implementation mechanisms will emerge from revisiting the social context and embeddedness dimensions to anticipate obstacles for any proposed option and then returning to the options menu to reassess the original choice. Such a back-and-forth activity may happen many times.

This revisiting process will also introduce the *trait-taking/trait-making* perspective and it will involve the *stakeholders* who are expected to be central to successful policy implementation. When this is done it should even include going back to the original definition of the problem to see if it still holds up under the series of analyses that have occurred, and then walking through the entire process again with some of the stakeholders.

Taken together, these various categories suggest a process for planning an excursion into the domain of natural resource policy reform. Given the hurdles we can expect to encounter during the implementation process, the policy and its implementation strategy need to be designed with both the intended behavioral changes and a knowledge of the terrain along the route in mind. This need can be met by following the process noted above.

That process is incorporated into a checklist called EPIC, for "Environmental Policy Implementation Checklist." The checklist is attached to the end of this volume as Appendix A, which can be used as a tool for applying the context map. It also includes the need to develop monitoring systems and contingency plans so that progress can be charted, strategies can evolve, and implementation can be redirected if outcomes and impacts deviate from intended courses.

Our odyssey from homogeneity through the world of context, then, has taken us from objectives and consequences, to policy clusters and mechanisms, to problem contexts and social contexts, through traits and stakeholders, and finally to a context map augmented by a process and a checklist. Along the way, we tested the conceptual framework by first overlaying it on a collection of African experiences and then on an Asian case study to see if it had any explanatory power. Finding that it did, we then retraced our steps and gained an overview of the context map that can be used as an heuristic device to assist in the design of environmental policy implementation strategies.

Now the final task of this volume is to step back and see the implications of the context perspective in two areas. The first area focuses on institutional options for sustainable development. The second area focuses on what might be done to incorporate a contextual focus into the actions of environmental policy promoters, the curricula of academic programs, and the conduct of policy research.

7

CONCLUSION: IMPLICATIONS OF THE CONTEXT PARADIGM

Dinosaurs have captured the human imagination. Bones, tracks, eggs, feces, and impressions from the distant past warn us that beings larger, stronger and as widespread as humans have walked the earth and then disappeared. Their fate grips our fears.

Much effort has gone into explaining why and how the dinosaurs became extinct. Some scholars have blamed shifting continents and climate change. Others have pinpointed volcanic eruption as the cause. Still others have cast a comet or meteor in the culprit's role. Many of those seeking the cause of the dinosaurs' demise have argued passionately for one factor or another as the key agent of mass destruction. They have joined a heated debate over competing theories of the great trigger.

The accepted explanation today is that the primary cause was the impact of an asteroid that plunged into what are now the Gulf of Mexico and the Yucatan Peninsula. The presence of iridium in sedimentary rock about sixty-five million years ago and the discovery of remains of the suspect crater suggest that this event set the great mass extinction in motion and created the opportunity for the rise of the mammals and the drama of humanity's ascendance (Alvarez, 1997). We owe our existence to a combination of a cosmic occurrence and its survival by small animals wearing fur and producing milk.

Although it probably is the correct interpretation of the original cause, the asteroid impact does not explain how some species persisted in different regions of the globe, including our mammalian forbears. Even those who assembled the evidence for the impact scenario recognize that explaining why all terrestrial life was not extinguished requires more understanding of the course of events and the role of various mechanisms that may have protected some species in different locations. In fact, some researchers adhere to a fuzzier and less dramatic perspective that resists single-cause explanations. They see a convergence of interacting events as a more accurate interpretation of an incomplete record. No one factor was totally responsible. This view is represented by a marvelous book on past and present extinctions by Peter Ward (1994) and by a statement made by Karen Chin, of the Museum of the Rockies, who said, "I think it was a combination of factors. There probably was an asteroid impact sixty-five million years ago. But I'm of the its-probably-more-complex-than-we-want-it-to-be school of thought." (quoted in Psihoyos, 1994: 255)

The demise of dinosaur dominance on this planet cleared the way for the rise of our own species, *Homo Sapiens*. For one hundred years after Charles Darwin's articulation of natural selection the central focus of human evolutionary studies was on the change in hominid characteristics themselves—the fossil remains of our precursors were considered far more exciting, and far more important, than the remains of other creatures found nearby. A skull, a skeleton, a tool, or a tooth generated more interest than the seeds, hippo bones or pollen traces found with them. But this has changed. Embracing an ecological approach to paleontology has revolutionized the discipline and supported the spread of a pan-disciplinary perspective called paleo-ecology. Today field researchers are as interested in documenting all the fossil animals and plants found with humanoid remains as they are in the remains themselves because this information makes it possible to reconstruct the environmental context of human development. As one observer noted,

Some adherents of this approach went so far as to say that once you had collected the fossil or artifact and had its context established (both in terms of time and paleoecology), you could throw the specimen away and still have 90% of its scientific worth. No one would advocate such a heinous act; it was just a dramatic way to make the point that contextual data had become very important. (Boaz, 1997: 150)

This statement parallels the school of thought about environmental policy implementation that emerges from this book—it is not simply the policy event that produces success or failure. Rather, it is the fit between events and settings that produces outcomes—the same event in different circumstances can produce different consequences. And knowledge of differences in circumstances can be key to understanding what happens, why it happens, and how to make desirable things happen when we need to.

This final chapter explores the implications of this context paradigm by addressing three pertinent issues. The first involves a current debate about the relative advantages of different institutions worldwide, such as the World Bank vs. non-governmental organizations (NGOs), as the lead players in efforts to achieve environmentally sustainable international development. The second deals with the promises and limitations of policy in guiding environmental change. And the third identifies specific lessons that apply to efforts to forge a sustainable future. Each issue is discussed below.

ORGANIZATION THEORY AND INSTITUTIONAL STRATEGIES

There are three competing hypotheses about the design of institutions for environmental protection and sustainable development. First is the quality hypothesis, second is the autonomy hypothesis, and third is the context hypothesis.

The Quality Hypothesis

This hypothesis postulates that it is the level of technical expertise embodied in an organization that will determine its effectiveness at adapting to changing needs. Ideological proclivity, empathy for a cause, personal commitment and a history of positive action are all less important than sheer competence.

This view is founded in the work of Clarke and McCool (1985) and others. Their work suggests that it is the expert organizations, such as the United States Army Corps of Engineers or the National Irrigation Administration of the Philippines, that have the most potential for achieving environmental objectives—not necessarily the organizations with the proper ideological orientation. Indeed, the untapped potential of military organizations in a depolarized world becomes a major focus, and bureaucratic reorientation becomes the strategy focus. Globally, this logic supports a technocratic approach to environmental policy that places the World Bank and the Global Environmental Facility (GEF) in the apex position among institutions. It expects the frontier of environmental accounting and investment to be pushed forward by the Bank and its staff, and, indeed, there is evidence that this has happened (Ahmad, Serafy and Lutz, 1989).

Such a view confirms the beliefs of natural scientists who trace the poor quality of environmental policy to the limited participation of ecological scientists in the policy-making process. It parallels the "getting the prices right" crusade with a "getting the knowledge right" initiative as the key to sustainable economies. It posits good science as the foundation for good policy.

The poor performance of many third world nations since independence offers justification for this view. Sluggish economic accomplishments, dim records of human rights and democracy, and dismal environmental protection are depicted as the result of misdirected leadership and policies. In this view, "environmental adjustment" programs would follow the path forged by economic adjustment efforts.

From this perspective, institutional redesign is equated with getting the right kind of competence in the driver's seat. And it has very clear strategic implications—it is a technocratic manifestation of command and control that leads to a proliferation of environmental regulatory agencies worldwide, to the setting aside of vast biosphere reserves where human impact is minimized, to the adoption of national policies consistent with the World Conservation Strategy, and to a reorientation and strengthening of the World Bank as the leader of the charge against unsustainable development in poor countries. Armed with the GEF and the power of the global financial institutions, it would impose a new environmental order. It would unseat economists and replace them with ecologists as the purveyors of the dominant paradigm, as the masters of the agency budgets, and as the voices of reason and right.

But the World Bank, the Army Corps of Engineers and this model also have been cast as villain, not hero (Rich, 1994). Critics blame the production-oriented, re-engineer the planet, and move-the-money-at-any-cost mind set for great expanses of ecological destruction. The obsession with large scale activity, the desire to alter physical landscapes, the market-based blinders borrowed from neo-classical economics, and the institutional arrogance all add up to an attack on the environment, not a defense of it. There is fear that even ecologists can not aim such a technocratic bulldozer in the right direction.

Low confidence in the ability of a combination of concentrated competence and fiscal might has led to a very different institutional prescription. This prescription is based on the autonomy hypothesis.

The Autonomy Hypothesis

The autonomy viewpoint postulates that self-managing small workgroups characterize the organizations and communities of the future. Here the foundation for creative response is the team approach that is found in NGOs, in some third world villages and in industrial organizations that recently have been created or redesigned based on autonomous workgroups instead of rigid hierarchies.

This hypothesis reflects the writings of Peters and Waterman (1983), Tomasko (1993) and others in the industrialized world and emerging nation observers such as Harrison (1987) and Rich (1994). Experience ranging from the decline of IBM as an industry-dominating giant to the dissolution of the Soviet Union can be marshaled to support this position. Its logic supports raising such organizations as Appropriate Technology International, the African Development Foundation and the Inter-American Development Foundation to positions as the prime movers in sustainable development in the third world, relying on citizen-neighborhood-community groups in the industrial nations, redistributing power from governmental to non-governmental institutions and gatherings, and lowering funding for the multilateral banks, major bilateral aid agencies, and national governments. It implies a massive shift of effort.

This perspective marks the ascendance of E. F. Schumacher's "Small is Beautiful" world view (1973). The model building block for a world society, in this paradigm, is much like the Amish community. The dream is to eschew consumerism for a simple life with a small ecological footprint. In fact, advocates of this perspective decry the internationalization of markets and the drive for free trade (Mander and Goldsmith, 1996). Stakeholder self-management is the preferred policy strategy—just as the quality perspective leans toward command and control as a preferred policy cluster, the autonomy hypothesis leans toward stakeholder self-management as the panacea for environmental deterioration. Indeed, those who promote this approach often argue that this is the *only* approach that will work over the long run.

This argument has four things in its favor. First, its romantic appeal is well received in an historical period marked by the dismantling of government machinery in places as different as South Africa, the former Soviet Union, and the United States. Second, there is evidence that variations on this theme can work under a wide range of conditions (Western and Wright, 1994). Third, it fits comfortably into a wave of downsizing government budgets. And, fourth a generation of reformers that tried to reorient bureaucracies has tired and would welcome a shift to smaller, and more optimistic, venues.

The autonomy viewpoint is in marked contrast to the quality perspective. Quality lends itself much more readily to an elitist, technical and purely scientific orientation. Autonomy, however, implies that a flat organizational profile, social proximity, and non-hierarchical approaches are more likely to produce viable resource management strategies than the imposition of techni-cally-correct criteria. This finds an eager following at a time when anti-exper-tise, volunteerism, and amateurism are equated with political liberation and when scientific illiteracy is widespread (Sagan, 1995). Indeed, autonomy is politically much more radical than quality. It appeals to disenfranchised, im-poverished and peripheral populations. And it can match the quality argu-ment in the attempt to capture the moral high ground and cloak itself in self-righteousness.

But it, too, has been criticized. Non-governmental environmental groups, for example, have been accused of conducting a "siege on science" (Fumento, 1993). Because of their limited scientific expertise combined with their de-pendence on fund-raising for survival, advocacy groups have distorted facts about environmental threats to raise salience, obtain publicity, and increase contributions to their coffers. Often staffed mainly by lawyers, they have pursued advocacy at the expense of science, both natural and social.[1] And with their entry into the realm of credit cards, mass mailings and marketing of logo-carrying items, they may have left their espoused missions even fur-ther behind. This may have resulted in the adoption of some scientifically unsound, economically unviable, wasteful, attention-diverting, and socially questionable policies (Bonner, 1993).

Indeed, anyone who has dealt with environmental or community-based groups in either developing or developed countries knows that their organi-zational capacity and their leaders' and members' individual competence ex-hibit chaotic patterns. A few are quite good and secure, most are quite weak and vulnerable. The international NGOs also vary considerably in organiza-tional prowess, financial support, technical strength, and political savvy. There is often an imbalance between technical and organization/process skills—those with either may lack the other. And some organizations truly represent the interests and perspectives of local communities, whereas others are little more than gentle, or not so gentle, predators (Josiah, 1996).

Those who see the future in terms of small-scale communities controlling pockets of nature dismiss the imperfections. They believe, period. But this study, with its situational and behavioral perspective, suggests another, but less righteous, alternative that does not exclude either of those above. That is the context hypothesis.

The Context Hypothesis

The context perspective has been set forth in this book. We need a combination of institutional avenues to address the diversity of situations surrounding natural resources on this planet. Both the autonomy and quality paradigms are needed. Both NGOs and the GEF have important roles to play; governments and the private sector both are parts of the solutions, and both small and large scale initiatives will be included in successful strategies to achieve sustainable development. But no one approach is adequate alone. And neither the technical proficiency of the World Bank nor the social propinquity of NGOs are adequate to address the range of circumstances and needs in the realm of environmental protection and sustainable use. Nor is the devolution of resource control to local communities a panacea for all time and all places.

The context hypothesis has no favorite policy cluster. Each may be more or less effective depending on local circumstances. Indeed, failure to consider the fit between each cluster (and combination thereof) and the social and problem contexts will lead to mistakes. Options should not be chosen without contextual assessment. Solutions cannot be determined a priori.

But there is a place where the context hypothesis departs from the quality and autonomy hypotheses. Quality and autonomy both focus on the vertical axis of Figure 3–A, whereas context includes both the vertical and horizontal axes within its cast. Indirect, seemingly unrelated, policies can produce webs of incentives and disincentives that overwhelm, and even negate, the direct policies. Thus, a context perspective leads toward a multiple policy orientation rather than a preferred solution.

National policies affect community robustness. Community health affects national and international well-being. The arrows go in both directions. Resources cross political boundaries. Resource threats cross political boundaries. Situations vary and the responses must also vary. That means alternative institutions must be in place to offer a range of responses to problems.

For example, the World Bank, GEF, and United Nations are poised to deal with cross-border, global commons, and regional policy problems. But as facilitators of community management schemes they are sorely equipped. Indeed, their overbearing management approaches and financial reporting requirements can get in the way of effective community operations. Self-management assistance lies more within the domain of the African Wildlife Foundation, Conservation International, World Wildlife Fund, and similar organizations. But alliances of funding, advocacy, and implementing organizations may be needed to create supportive settings.

Likewise, in the United States, state-level initiatives and policy reforms are needed to complement national exercises such as the President's Council on Sustainable Development. Community-based efforts also must go forward. There is no single arena where the problems can be solved. And no one approach will work in all settings, from the Mississippi basin to the Harlem River to the Denali Mountain to the Hawaiian coral to the Olympic Peninsula to the Rio Grande to the Everglades to the Appalachian hinterland to the Chesapeake Bay to the Connecticut River—urban and rural, high and low, rich and poor, homogenous and heterogeneous, dense and sparse, hot and cold—the variety is hard to imagine. The responses must match that variety. Otherwise, unanticipated impacts will result from attempts to impose uniform rules on these settings. And some of the major determinants of those impacts most likely will come from indirect sources.

Indeed, indirect incentives exhibiting low insulation will negate many direct policy thrusts, not just in the United States, but world wide. Thus, the forces directing international financial flows and the actors engaged in environmental protection need to work in concert. The World Bank needs to be reoriented, not replaced. Its large scale focus could be redirected toward restoring degraded environments and its financial leverage could be used to help turn around national policies that contribute to degradation. The International Monetary Fund, then, would also need greening to support environmental adjustment. But adjustment of this type would exhibit diverse policy approaches to match the problem and social contexts of local situations.

One aspect of context, for example, is informational openness, which is linked to human rights and democratic processes. Bilateral donors, such as the Dutch and Americans, are presently concerned with this issue. But it should be more closely linked to community resource management policies. NGO-bilateral partnerships could help to move national social contexts in a direction supportive of effective environmental management. But the link needs to be made clear.

Specific cases, however, will demand different responses. For example, different degrees of embeddedness combined with different resource-threat relationships may dictate some bizarre institutional mixes. Until a context analysis occurs, however, the appropriate response is unknown. The obvious and intuitive may be counterproductive. Context inserts itself in subtle and counterintuitive ways.

This emphasis on context is a challenge to some analytical tools, academic disciplines, and professional perspectives that claim universal suitability. It suggests that, like some good wines, all things do not travel from place to place equally well. And probably none can perform under all conditions. Even so, if diversity is to be addressed seriously, many bureaucracies will yield to combinations of networks and temporary task forces that are assembled around diverse contexts. Classical economic mind sets will be balanced by others. *Ecological Economics, Ecological Anthropology,* and *Conservation Biology* are harbingers of fields to come. And all three, if successful, will blur the boundaries of social and natural science, and in so doing transform policy science and organization theory.

Many of the problems associated with the quality and autonomy hypotheses could have been predicted from traditional organizational studies. Within environmental advocacy groups, for example, the need to extract resources from their environments would lead to the displacement of their original goals. And studies of expert organizations would predict the increasing insularity and arrogance of an organizational culture such as that of the World Bank. Thus existing organization theory (Thompson, 1967) helps to explain some of the institutional problems we face today. Unfortunately, however, it does not give us what we need to make the changes because it lacks different models of context.

Context goes beyond geography and academic disciplines. It also has a temporal dimension. Ptolemy's analysis of celestial movements was considered accurate explanation before the Copernican revolution. But the displacement of the Ptolemaic world view was only achieved in the face of intense institutional resistance. Indeed, organizations and institutions are themselves contextual creations—organizations strongly reflect the times that generate them (Stinchcombe, 1965) and so do development strategies.

The multilateral, bilateral, and non-governmental development institutions of today are no exception. They were innovative reactions to the cold war atmosphere of the immediate post World War II era. Their monolithic natures, their utilization and enhancement of civil engineering and neo-classical economic tools and perspectives, their rigid hierarchies and technocratic cultures, and their opaque organizational styles are no accident. Cold war origins made such attributes natural. The imperatives of the period dictated just such institutional responses.

Since then, however, the world has changed. Not only has the collapse of the Soviet empire shattered tri-modal definitions of the political world and inserted diversity as a newly discovered virtue, but the focus on extra-governmental forms of human organization and the revolution in communication technology have combined to create a climate that enables (and legitimizes) decentralized, transparent modes of collective action. Organizations created today will be no less prisoners of temporal context than those created half a century ago in the wake of World War II. NGO networks, task-force based organizations and scrutable decision processes fit the present, and the need to reform or replace vestigial donor and national institutions is real.

Indeed, the call for a new Bretton Woods conference to redesign the array of international institutions such as the World Bank group, International Monetary Fund, and General Agreements on Tariffs and Trade (GATT) has merit—these arrangements were based on both political realities and conceptual foundations that are dubious today. But we should not lull ourselves into believing we will solve institutional problems for all time. Many of our answers will probably have half lives of no more than three decades. Uncertainty argues for flexibility and an array of available alternatives—a contextual strategy.

IMPERFECT KNOWLEDGE
AND COMBINATIONAL CAUSALITY

Natural resource and environmental policies, as noted earlier, are not really aimed at nature—nature has little truck with the proclamations of this single species or its institutions. Such policies actually aim at human behavior. But human behavior may not always be identifiable, with a reasonable degree of certainty, as the major culprit in resource degradation. We may be the source of the problem (Ward, 1997), and then again we may not be. And policies that precede knowledge may act to exacerbate situations. Although it is fashionable to put humankind at center stage when it comes to blame (McKibben, 1992), we need to be approximately right about why things happen. It is far better to be approximately right than to be precisely wrong.

As we argue in this volume, for policy to work it must be aimed explicitly at human behavior. But hubris results from thinking that all problems are amenable to policy solutions. Even though the impact of human actions on the biosphere has certainly catapulted the policy sciences into prominence, some things are simply beyond the reach of policy.

Beyond Policy

We are reminded of the limits of policy by Christopher Stone, who warns us about rushing to judgment about causes of environmental change:

. . . periodically the oceans are marked by "blooms"—sudden swellings in oceanic algae populations. These episodes have lethal effects on marine life . . ., either through depriving other marine life of adequate oxygen or, in the case of "red "tides," poisoning them through associated toxins. There is considerable concern that the algae blooms are fostered by man-made pollution. . . . On the other hand, [they] have been reported periodically through human history . . . long before our tampering with the environment had reached significant levels.

He then goes on to point out that, in addition to the scientific difficulties, there is a political dimension to problem definition:

. . . even if human activities . . . are exacerbating the algae blooms, there is considerable room for uncertainty as to what chemical agents and which nations' activities are causally responsible. Nation A may point the finger at nitrogen emanating from nation B's agriculture; nation B blames phosphorous and trace metals from nation A's industrial waste—and there may well be a grain of truth to each position.

And he further expounds on the limits of human action in the face of non-linear change and truly universal processes:

. . . while human activity may be affecting climate change in significant ways, the principal determinants of global climate remain natural forces [such as] solar radiation, the tilt of the earth's axis with respect to the elliptic, and the eccentricity of the earth's orbit. . . . There are enormous and wide-ranging climatic effects of such other natural processes as solar activity, volcanic debris, and the El Niño, as well. All these "background" processes complicate efforts to determine the extent to which any perceived changes in the environment can be pinned on human activity—and therefore be subject to mitigation by changes in human conduct, through law. [and policy, we would add] (Stone, 1993: 105–6)

We must remember, first, that we are only one of the many players on the environmental history stage; and second, that seldom can effects be reduced to single causes. This study suggests that combinations of policy substance and implementation decisions interacting with contextual conditions will be the heros and culprits in any environmental policy drama. The search for simple answers is itself part of the problem. There is not only no silver bullet, there is no single target to aim for.

At the same time, paralysis is no answer. Learning requires a combination of observation and action. Both occur through institutions, and there is no lack of prescriptions for redesigning the institutional matrix of international development and environmental protection. But prescriptions often miss the mark. This is especially the case when they are based on ideology.

Beyond Ideology

As we have seen, there are three major policy-based approaches to solving environmental ills. The first is regulation—the application of command and control mechanisms to constrain or prohibit certain types of human behavior. Most traditional protectionist strategies toward natural areas and preventative approaches toward pollution reflect this option. Indeed, environmental protection is often equated with command and control. Some environmental advocates see this as their most effective tack, and even elevate it to the status of philosophical departure. And many opponents of environmental protection use this to their advantage by depicting environmentalists as control freaks who are out to impose a warped world view.

The second alternative—stakeholder self-management—also has its following. Those who espouse libertarian approaches to government, those who place a high value on small-scale community as the building block for society, and those who consider the protection of threatened cultures as a priority, all coalesce around this as a preferred alternative, and even a panacea. It is represented by a literature that is growing daily. Indeed, it has been elevated by some to the level of ideology. And those who hope to thwart environmental initiatives sometimes cloak themselves in this rhetoric to hide their real intentions while they privatize rights over resources and proceed to destroy them in the name of liberty and justice.

Between these two extremes lies the third alternative—incentives to reward and guide action, but neither imposing controls on behavior nor devolving control over all decisions about the use of a resource. Those who advocate market mechanisms to deal with social problems in general embrace this approach. This is the foundation for market-based environmentalism. Like the others above, this approach also assumes ideological proportions among market maniacs. And, like the others, it harbors those who wish to reward the short-term mining of nature's bounty.

But the context hypothesis suggests that all of the above have roles to play, and that the focus should extend past direct strategies to encompass indirect incentives as well. No one option should be embraced in exclusion of the others. The most effective policy alternative(s) will be determined by a combination of contextual factors that include the problem context, the social context, the degree of embeddedness and the behavioral objectives of the policy. Ideology should not dictate the choice. And ad hominem arguments should not drive the decision calculus.

A regrettable example of this appeared in a very respectable journal. The author questioned any positions emanating from business sources or any initiatives with business people involved with them that used the term "sustainable development." (Willers, 1994) He equated this to "green washing"—a deception to allow manipulation of the media so that profiteering could continue at the expense of the environment and with the acquiescence of the public. This was not a substantive contribution to the elaboration of sustainable development. It did, however, represent the danger of ad hominem and ideological interpretations of the world.

This is especially important due to three things. First, the discussion of salience found that both false and factual information have assisted policy implementation by raising salience—the truth was less important than the urgency of the issue. This lends itself to ideological manipulation and it could lead to consequences way out of the expected realm. While recognizing the importance of salience for inducing change, we also need to foster cooperation in discovering truth. And we need to accept that truth does not always support our preferences, nor does it remain static. Today's facts may become tomorrow's myths, and these myths may become barriers to implementing further changes.

Second, the discussion of systemic objectives alerted us to the fact that sustainable economies will exhibit new ways of measuring what we do. This implies changes in organization, changes in accounting and accountability, and changes in the ways we earn our livelihoods, and it suggests that *business*, and how it is conducted, will be central to the shift, not peripheral to it (Cairncross, 1992; Hawken, 1993).

This means that some of the traditional "bad guys" of the environmental drama must be included in the team of "good guys." There is evidence that this is already happening, but it will need to go further—holier-than-thou environmentalists will need to link arms with such organizations as the World Bank, Army Corps of Engineers, and defense industry giants to create ecologically sound organizational missions and to build a sustainable economic system. And it means that "hard nosed" business leaders will recognize that the "bottom line" of today's accounting methods only goes half-way down the page. Business executives will become stronger, and their organizations will become more viable, by inventing the economy of the future in tandem with the environmental community. Community bankers will use new criteria for assessing investment potential and making loans to entrepreneurs. The future of business is intricately tied to the resilience of the biophysical environment, and the measurement of profit and performance must reflect these ties. But for this to come about, ideological camps and ad hominem rhetoric must yield to new alliances that focus disparate perspectives and diverse types of expertise on the common problem—survival.

Third, we have seen the limitations on our knowledge. Uncertainty and the limitations of policy add to the need to consider a wide range of alternatives, experiments and innovative approaches custom-tailored to different circumstances. Good ideas may come from strange quarters. No individual, group, profession or institution has a monopoly on truth. We are all in this together. Ideologies of any type limit the range of options and can derail sustainable initiatives simply because those initiatives may lie outside the area of ideological acceptability or the source of legitimacy. The cost of this could be enormous.

In the United States, for example, a form of intellectual blindness has emerged. It is called "political correctness." The attempt to dictate what conglomeration of political perspectives and language uses constitutes truth is creating a barrier to understanding and resolving social issues. Indeed, it legitimizes ideology as the filter for seeing the world and it makes the generation of knowledge and the critique of behavior more difficult. It even harnesses the term "diversity" to its cause and distorts the meaning of the term by focusing not on the variation among settings but rather on the racial characteristics of people. The context hypothesis supports the need to advance far beyond political correctness.

The context hypothesis, then, promotes a pragmatic view based on situational diversity. It encourages solution packages custom tailored through time, space, and substance. It rejects ideology as the basis for environmental policy decisions. It suggests that good social science must examine the fit between context and action. And it also offers the possibility that disaggregating behavioral objectives and attributes of problem settings can improve our policy strategies and increase the chances that we will be able to create a sustainable future.

DESIGNING SUSTAINABLE FUTURES

Recent studies by natural scientists conclude on grand themes, such as protecting biological diversity and fostering sustainable development (Ward, 1994; Wilson, 1994). But they give little, or no, insight on *how* to do it. And such insight is needed for two reasons. First, although our own species may not be the only cause of environmental change, we *are* one of the causes—some suggest we are changing the evolution of life forms on this planet (Ward, 1994; Weiner, 1994), and others suggest we are even influencing the planet's spin through the construction of large water reservoirs in the northern hemisphere. Second, the speed of ecological change appears to be rapid and we cannot wait for future generations to compensate for our accumulated errors—we need swift action to alter our behavior. Although we will ultimately be judged as a species on the geological time frame, we need to move in a management time frame. Thus, we need specifics.

We have already taken some steps beyond the grand themes. This study has made some specific suggestions regarding how to develop environmental policies to fit the settings they will engage. Chapter Six and Appendix A present a process and a checklist for those contemplating policy action. Now we will add some recommendations that flow from the study. Aimed most directly at international organizations, they also are applicable to domestic public and academic institutions in the United States and elsewhere. These suggestions are organized under behavioral, program, and research headings.

The Behavioral Imperative

Our analysis of different intended consequences in Chapter Two brought human behavior to center stage. We showed that different objectives require different means. And Chapter Four itemized categories of context that channel behavior. The conclusion flowing from that exercise is that

- the departure point for natural resource policy design should be changing human behavior.

To identify appropriate behaviors it is first necessary to comprehend the ongoing situation. That means that

- the context of environmental problems should be analyzed to determine the complex of incentives that guides the human behavior that contributes to them; and

- the degree and nature of embeddedness should be determined, and program components should be designed accordingly, in order to facilitate the transition from an undesirable embedded situation to a preferred set of behaviors.

This behavioral focus is an important departure from present approaches because it goes to the reasons why people are doing what they are doing. It does not just superimpose a new rule on a system, but rather it attempts to understand the system and redirect its dynamics. The assumption is that there are reasons why people do what they do and, until those reasons are made explicit, effective policies cannot be designed.

If behavior, and its context, does become central to policy design, then there are implications for the ways that international institutions conduct business. Some of the implications apply to the programs they fund and some apply to their internal operations.

Program Departures

International donor and national government strategies need to precede policy initiatives with natural resource sector studies. Such studies do occur. But they should be much broader than current practice. Indeed, this report suggests that

- any natural resource focused sector analysis should spread out from the resource itself to encompass the full range of human behavior affecting it, to understand the reasons for that behavior, and to examine all four policy clusters to extract the role of policy in promoting such behavior.

Especially indirect incentives, which by their nature will be external to a traditional sector focus, must be examined. The resource offers a beginning point, not a closed boundary, for a sector study. Any human behavior that affects a natural resource lies within the realm of the analysis. The analyst will act as a sleuth, tracing long linkage trails from policy options, to human behavior, to natural system consequences. Such detective work will involve those who understand the local social context and live within it.

Casting this wider net is more apt to identify combinations of constraints and opportunities that add up to a singular situation.

- Project, program and policy design, then, will flow from an analysis that isolates those unique conditions that favor success.

Indeed, identifying unusual combinations of circumstances has already been associated with some aspects of success in Africa. For example, an evaluation of the Land Conservation and Range Management Project (LCRD) in Lesotho concluded:

Circumstances favorable to a project's success may also argue against its replicability. The location of the RMA [Range Management Association] was a deliberate choice based on a unique combination of favorable political, ecological, and functional factors. Because this combination of factors is unlikely to exist elsewhere in Lesotho, LCRD may be only partially replicable in other parts of the country. (Warren and others, 1985: 16)

This assessment doubted the ability to replicate a project *within* a small country, let alone *between* countries, regions, or continents. Over a decade later it is time to heed the warning. Successful reforms are more likely to exploit and build upon unique physical, historical or social attributes than they are likely to result from the imposition of a uniform organization, policy, or technology.

An emphasis on context should also lead funding organizations to

- abandon replication of form or substance as a central operating tenet and replace it with a focus on the fit between the context and the policy or program intervention.

The context should lead the analysis. The EPIC approach introduced here might be one experimental way of making this shift.

EPIC, in fact, provides a mechanism for disaggregating the dimensions of context to allow a custom-tailored policy strategy. Some of those dimensions could play key roles in developing the configuration of an implementation effort. For example,

- Project and program funding levels should be based on an intended interorganizational power balance.

Analysis of the budgets (including percent devoted to the wage bill), organizational linkages, and staff capabilities of both potential allies and competitors should become routine. When insulation is low the relative resources of environmental policy implementors will need to be disproportionately large. The context map offers guidance here.

Likewise, the context map supports the present emphasis on democratization. But the link between human rights and natural resource use should become more central. Voter behavior is *not* a central issue, but

■ support for a free press with competing views, combined with the training of competent environmental journalists and the promotion of ecological and scientific literacy, should become major elements of democratization efforts.

Indeed, informational openness needs to be elevated to a position of paramount importance. The global computer revolution combined with satellite technology and the collapse of some of the most opaque political systems provides an opportunity to alter context directly and create conditions to enable sustainable development.

■ Donors and international organizations should make open debate with free presses and transmitters a condition of assistance, and they should directly fund environmental journalism and scientific writing programs and the institutions to support them.[2]

But this need is not just found in developing countries. There also is a need to support scientific and environmental awareness on the part of the press corps in wealthy nations. The American national print and television media failed to cover the President's Council on Sustainable Development until the week before it issued its report. And local reporters fare no better—the media in Minnesota ignored the Sustainable Development Initiative and the launching of the Sustainable Development Roundtable. These initiatives were not understood, and since they aimed at collaboration rather than confrontation they were handicapped in the competition for headlines and sound bytes. Scientific and environmental illiteracy dominates the institutions that determine what information flows to the public. This needs to be rectified.

On the international scene, democratization should put less emphasis on voter behavior and more on open information generation and sharing within societies. And donors need to set an example. Thus,

■ democratization support programs need to be bolstered by transparent operating styles within the international organizations themselves.

The suggestion above brings us to the internal operations of the international donors and their program development processes. Programming is essentially the way a donor goes about conducting business. Since donors often act as wholesalers of services to third world entities, procurement policies are central to programming. One step to take, then, is to go beyond cold war categories of source and price. Thus,

- public institutions should transform themselves into buyers and users of environmentally responsible goods and services by requiring suppliers to demonstrate "green" production processes and service delivery mechanisms and practices.

Governments, donors, foundations, international organizations, and NGOs should support certification and monitoring efforts both among their own suppliers and as levers to assist worldwide markets to distinguish among different natural products and processes. Procurement policies and practices should adjust to the need to incorporate environmental soundness into business operations. And,

- the donor community should examine its own operating style to discover the environmental consequences of different modes of operation.

The effect of this would be to stimulate a market for methods to examine the environmental impact of different ways of doing business. Tools of full-cost pricing and life-cycle analysis should be applied to donor operations themselves. And both the tools and the operations would improve as a result. There is evidence that new communication technology is having an effect— electronic mail via satellite is lowering the need for travel by United Nations, World Bank, State Department, Agency for International Development (AID), conservation organization, consultant, and other private sector staff. But this is a de facto change resulting from cost considerations, not a conscious examination of the environmental consequences of different ways of operating. As long as budget objectives and ecological objectives coincide, this poses no problem. But this is not always likely to be the case. Conscious examination is needed at this stage in the drive for sustainable development.

Additionally, the internal organization of international donors and state or national government agencies might reflect new problem definitions. Regional divisions would most likely remain because the regions of the planet do contain major differences in terms of history and culture. And among natural resource management agencies a shift from a commodity-based organization to a landscape or ecological service orientation is beginning to appear. International organizations generally use a combination of geographic (such as Asia, Africa) and commodity (such as agriculture, water, forestry) sectors to organize their units. But subdivisions might be created to reflect different aspects of problem contexts—discreteness, non-linear progression of the problem, the different resource-threat relationships, and the degree of polity congruence found to hinder the implementation of solutions.

Moreover, both domestically and internationally,

- a recognition of the importance of indirect incentives argues for bringing into the fold many institutions that are presently not engaged in the search for sustainable societies—insurance, banking, real estate, military, and other institutions can pursue policies and engage in practices that contradict the drive for sustainable solutions—and it identifies the need for programs to educate them, to build new capacities within them, and to reorient them.

A context perspective, then, implies new program departures and it points toward new ways of conducting business and organizing to do it. But it also has implications for the way we develop the knowledge we use to do our business. The context hypothesis suggests a new way to focus applied policy research. Some of this builds upon current trends and some of it indicates a need for totally new departures.

Research Redefinition

The argument of this book is that policies and programs are co-actors with context. We can understand the results of policy initiatives only by charting how they fit into the settings they encounter. To do that, we need maps and typologies of settings and we need methods for contextual analysis.

The practice of international sustainable development has already been exposed to attempts to create contextual assessment methods. For example, participatory rural appraisal (PRA) and local eco-development approaches are evolving throughout the globe and some are rooted in context and adaptation. One PRA case study makes the point that

PRA is not a universal methodology. It must be adapted to local cultural, ecological, institutional and economic circumstances. This adaptation can only be done in the field, through collaboration among those who know the local circumstances and those who are familiar with PRA. (Ford and others, 1993: 43–44)

Such approaches are in the vanguard of the search for methods for contextual analysis—they are tied to people, place and history.

PRA is an important ex ante bundle of approaches that can be used in behavioral assessments and policy design. But ex post studies also need to incorporate participatory processes (Honadle and Cooper, 1990) and to follow through with the contextual theme. This implies that policy researchers should develop the maps and typologies noted above, and that

- policy and program evaluations will look specifically at the context-project fit as the independent variable in causal relations.

Scopes of work, evaluation studies, and the products of funded research will look quite different from most of those produced today. Both the findings and conclusions will contain new insights. The fit, not just the attributes, should define the independent variable set. Learning is not blind replication but neither is it ignoring other experience. The key is to

- fund studies that reorganize and reinterpret previously reported experience to capture the contextual element and learn from it.

These agenda should guide the inquiries of the such donor units as the evaluation division of AID's Center for Development Information and Evaluation (CDIE), the World Bank's Operations Evaluation Department (OED), and any successor units resulting from organizational re-engineering.

But applied research does not just happen as a result of international institutional activity. It also occupies the resources of universities and research institutes. Their endeavors should also reflect the changes proposed here.

- More general research should be fit-focused.

For example, research on the design of environmental protection institutions should look beyond organizational structure and process to the fit between the organizations and their contexts. This should be the norm, not the exception. Scopes of work for funded research should reflect this shift in emphasis and Ph.D.s in policy science and public administration should generate new contextual maps and reflect context-based research designs.

Four specific types of studies might further this transition. The first would examine the movements and evolution of organizational types (organographics?) just as population studies of humans and animals trace ups and downs, migrations and mutations. One study pioneered a similar approach (Hannan and Freeman, 1989) but it needs to be taken further to examine the worldwide diffusion of different organizational types and to focus on environmental management organizations, regulatory institutions, and local resource management groups. Indeed, ecological concepts such as "co-evolution," "sources and sinks," and "keystone communities" might be applied with success to organizational studies. The second type would use case studies to develop better understanding of organization-context dynamics. Some might be revisionist examinations of prior experience while others would be new experiences entirely. The third would conduct cross sectional comparative analyses of the cases. The fourth would be theoretical exercises to construct alternative context maps.

The recommendations above challenge policy makers and policy managers. But they also point to weaknesses in educational offerings. For example, environmental studies programs need to incorporate behavioral science knowledge and perspectives. That is because human behavior is key to ecosystem threats and environmental action. The short run demands more than the infusion of ecological literacy, biophilia perspectives, and environmental values—it needs professionals prepared to promote action to change behavior.

Likewise, programs of public administration and public policy education need to go in new directions. They reflect a preoccupation with the internal attributes, both processes and structures, of management organizations and the theoretical confines of neoclassical economics. It is time to get beyond organizational mechanisms and economic growth to engage both the human and natural contexts of policy implementation for sustainable development. It is also time to inject a comparative perspective into all management, policy and environmental studies programs, for it is only through comparison that the features of context come into view.

This study presents a new lens for viewing the links between human behavior, public policy and nature's bounty. It posits an underlying pattern to the way that context interacts with policy to affect outcomes. But it does not argue for a single natural "law of context." Rather, what it proposes is both more plausible and more adaptable than any simple, rigid set of relationships—it is the context hypothesis.

NOTES

1. For example, the spring, 1995, issue of *The Amicus Journal*, published by the Natural Resources Defense Council (NRDC), lists thirty-four lawyers and nine scientists as staff members of that organization. Thus the lobbying and litigating activities overshadow the science functions, at least in terms of the staffing pattern.

   But this does not show the whole picture. One of NRDC's nine scientists was elected a fellow of the American Physical Society and research by other NGO-based scientists is sometimes highly praised. For example, Dr. Theodora Colborn of the World Wildlife Fund is considered a pioneer in the identification of the impact of synthetic endocrine disrupters. Thus some of the work done by scientists in these organizations has been recognized as outstanding within the scientific community and their work is often published in professionally refereed journals.

   At the same time, Fumento and Bonner's points have merit. The issue of *The Amicus Journal* noted above offers an example—in response to an angry letter from a creationist, the editor backs away from supporting a very favorable review of Jonathan Weiner's excellent book *The Beak of the Finch*. Placating sources of finance appears to have the upper hand over endorsing solid scientific inquiry and its extension to a wider audience.

2. For an outstanding example of environmental journalism, see Yvonne Baskin's book, *The Work of Nature*, in the bibliography at the back of this volume.

EPILOGUE

I spent a few days working on an early draft of this volume in a cabin overlooking a freshwater lake in north-central Minnesota. As I glanced at the background of brilliant leaves on the trees across the lake and watched the loons diving, it reminded me of Malawi and my house along the northern shore of Lake Malawi twenty-five years before. There the scene encompassed the Livingstone Mountains of Tanzania as a backdrop to fishermen diving from their dugout canoes. And I remembered friends from that time—Kitha, who smiled at my reaction to the sunset; Sichone, a health worker at Kaporo; and Moses, an entrepreneur.

These men were my friends. Indeed, we taught each other many things. And after Malawi, as I wandered through the worlds of graduate school, international consulting, teaching and research, and public service, their images sat on my shoulder and I would often ask them "does this make sense?" Sometimes they laughed, and sometimes they nodded their heads in approval. They were an imaginary reality check.

All of them intuitively understood the importance of context. One had worked in the mines of South Africa and the others had been to Tanzania and Zambia. Each spoke various dialects and had experienced different places. They were living proof of a village-level appreciation for variety.

And, now with the introduction of new leadership to Malawi, they are experiencing contextual change in their remote homesteads. A new era has begun. And yet many of the global institutions promoting policies, programs, and particular definitions of problems seem determined to deny differences by standardizing strategies. Both environmental advocates and development engineers continue to cling to their own myths and models. And development administration and public policy researchers also seem determined to shine their lights on management tools and economic prescriptions. Solutions are still replicated and impacts are still attributed to the imported processes. The mesh between context and intervention is glossed over in the rush to peddle particular policy perspectives.

But now it is time to give context its due, and to recognize that thinking globally can lead to acting homogeneously. It is time to think locally, too. To succeed, policy change strategies must mesh with local realities. Whether or not the earth continues to be hospitable to us may be determined by our ability to guide our own behavior by matching strategy to context. The challenge is to act on this understanding.

It is easy, however, to succumb to gloom. A planetary human population of six billion, plummeting biological diversity and resource stocks, the urbanization of natural settings, the cascading ecological crises, the institutional shock waves traveling around the globe, and the bombardment with predictions of impending disaster, all combine to make cynicism and pessimism easy to embrace. And they, in turn, fuel a sense of ignorance and helplessness in the face of overwhelming forces, a fear of the future, and then withdrawal from public debate and disengagement from change efforts. But that is the easy course.

The more difficult, but also the more promising, course follows paths of opportunity. Although prospects sometimes look dim, these are exciting times—human accomplishment during the past fifty years has been unprecedented. The institutions we take for granted today are mostly young and they are departures from previous human experience. But frontiers have not disappeared. The drive to develop sustainable societies will produce even greater departures in ideas, institutions, technologies, and opportunities.

Contextual thinking and contextual action may take us down new paths that will help to delay the extinction of our species. Those who blaze these paths may discover new solutions for our problems and new opportunities for accomplishing great things. The real challenge, then, is to lead the charge and not to be swept away by it.

BIBLIOGRAPHY
AND INFORMATION SOURCES

In addition to personal work experience in twenty-seven countries over the last thirty years, and interviews with forty-six people in twenty organizations over a two-year period, this study builds upon information and perspectives contained in the following two hundred ninety-six books and book chapters, ninety-six articles in journals, proceedings, and periodicals, one hundred thirty-seven published and unpublished reports and other documents, and six videotapes.

BOOKS AND BOOK CHAPTERS

Abernathy, Virginia. 1993. *Population Politics: The Choices That Shape Our Future*, New York: Plenum Press.

Adams, W. M. 1990. *Green Development: Environment and Sustainability in the Third World*, London: Routledge.

Adams, W. M. 1993. *Wasting the Rain: Rivers, People and Planning in Africa*, Minneapolis: University of Minnesota Press.

Agarwal, Bina. 1986. *Cold Hearths and Barren Slopes: The Woodfuel Crisis in the Third World*, London: Zed Books.

Aiken, S. Robert and Colin H. Leigh. 1992. *Vanishing Rain Forests: The Ecological Transition in Malaysia*, Oxford: Clarendon Press.

Allison, Graham. 1975. "Implementation Analysis: The 'Missing Chapter' in Conventional Analysis—A Teaching Exercise," in Richard Zeckhauser, ed. *Benefit-Cost and Policy Analysis 1974*, Chicago: Aldine.

Anastas, Paul T. and John C. Warner. 1998. *Green Chemistry: Theory and Prac-*

tice, Oxford: Oxford University Press.

Alvarez, Walter. 1997. *T. Rex and the Crater of Doom*, New York: Vintage Books.

Amy, Douglas J. 1987. *The Politics of Environmental Mediation*, New York: Columbia University Press.

Anderson, Anthony B. ed. 1990. *Alternatives to Deforestation: Steps toward Sustainable Use of the Amazon Rain Forest*, New York: Columbia University Press.

Anderson, Robert and Walter Huber. 1988. *The Hour of the Fox: Tropical Forests, the World Bank and Indigenous People in Central India*, Seattle: University of Washington, DC Press.

Anderson, Terry and Donald Leal. 1991. *Free Market Environmentalism*, San Francisco: Pacific Research Institute.

Argyris, Chris and Donald A. Schon. 1974. *Theory in Practice*, San Francisco: Jossey-Bass.

Armitage, Jane and Gunter Schramm. 1989. "Managing the Supply of and Demand for Fuelwood in Africa," in Gunter Schramm and Jeremy Warford, eds. *Environmental Management and Economic Development*, Baltimore: Johns Hopkins Press, pp. 139–71.

Ascher, William and Robert Healy. 1990. *Natural Resource Policymaking in Developing Countries: Environment, Economic Growth and Income Distribution*, Durham: Duke University Press.

Asmerom, H. K. and R. B. Jain, eds. 1993. *Politics, Administration and Public Policy in Developing Countries: Examples from Africa, Asia and Latin America*, Amsterdam: VU University Press.

Atlhopheng, Julius, Ghadzimula Molebatsi, Elisha Toteng and Otlogetswe Totolo. 1998. *Environmental Issues in Botswana: A Handbook*, Gaborone: Lightbooks.

Bagadion, Benjamin. 1988. "The Evolution of the Policy Context: An Historical Overview," in F. Korten and R. Siy, eds. *Transforming a Bureaucracy: The Experience of the Philippine National Irrigation Administration*, West Hartford: Kumarian Press, pp. 1–19.

Baker, Randall, ed. 1992. *Public Administration in Small and Island States*, West Hartford: Kumarian Press.

Bartlett, Robert V. 1994. "Evaluating Environmental Policy Success and Failure," in Norman J. Vig and Michael E. Craft, eds. *Environmental Policy in the 1990s*, second edition, Washington, DC: Congressional Quarterly Press, pp. 167–88.

Barzetti, Valerie and Yanina Rovinski, eds. 1992. *Toward a Green Central America:*

Integrating Conservation and Development, West Hartford: Kumarian Press.

Baskin, Yvonne. 1998. *The Work of Nature: How the Diversity of Life Sustains Us*, Covelo, Calif.: Island Press.

Bates, Robert. 1981. *Markets and States in Tropical Africa: The Political Basis of Agricultural Policies, Berkeley*: University of California Press.

Beatley, Timothy and Kristy Manning. 1997. *The Ecology of Place: Planning for Environment, Economy, and Community*, Covelo, Calif.: Island Press.

Bergesen, Helge Ole and George Parmann, eds.1993. *Green Globe Yearbook of International Co-operation on Environment and Development*, Oxford: Oxford University Press.

Berry, Wendell. 1981. *The Gift of Good Land*, San Francisco: North Point Press.

Bingham, Gail. 1986. *Resolving Environmental Disputes: A Decade of Experience*, Washington, DC: The Conservation Foundation.

Blaikie, Piers and Harold Brookfield. 1987. *Land Degradation and Society*, New York: Methuen.

Boaz, Noel T. 1997. *Eco Homo: How the Human Being Emerged from the Cataclysmic History of the Earth*, New York: Basic Books.

Bonner, Raymond. 1993. *At the Hand of Man: Peril and Hope for Africa's Wildlife*, New York: Knopf.

Botkin, Daniel B. 1990. *Discordant Harmonies: A New Ecology for the Twenty-First Century*, New York: Oxford University Press.

Broad, Robin with John Cavenagh. 1993. *Plundering Paradise: The Struggle for the Environment in the Philippines*, Berkeley: University of California Press.

Bromley, Daniel W., ed. 1992. *Making the Commons Work: Theory, Practice and Policy*, San Francisco: Institute for Contemporary Studies.

Brown, Lester et al. 1991. *Saving the Planet: How to Shape an Environmentally Sustainable Global Economy*, New York: W. W. Norton & Company.

Browne, William, Jerry Skates, Louis Swanson, Paul Thompson and Laurian Unnevehr. 1992. *Sacred Cows and Hot Potatoes: Agrarian Myths in Agriculture*, Boulder: Westview Press.

Bryson, Reid and Thomas Murray. 1977. *Climates of Hunger: Mankind and the World's Changing Weather*, Madison: University of Wisconsin Press.

Caiden, Naomi and Aaron Wildavsky. 1974. *Planning and Budgeting in Poor Countries*, New York: John Wiley & Sons.

Cairncross, Frances. 1992. *Costing the Earth: The Challenge for Governments, the Opportunities for Business*, Cambridge, Mass.: Harvard Business School Press.

Callenbach, Ernest, Fritjof Capra, Lenore Goldman, Rudiger Lutz and Sandra Marburg. 1993. *Ecomanagement: The Elmwood Guide to Ecological Auditing and Sustainable Business*, San Francisco: Berret-Koehler Publishers.

Carroll, Thomas F. 1992. *Intermediary NGOs: The Supporting Link in Grassroots Development*, West Hartford: Kumarian Press.

Catton, William R., Jr. 1982. *Overshoot: The Ecological Basis of Revolutionary Change*, Urbana: University of Illinois Press.

Cernea, Michael. 1985. "Alternative Units of Social Organization Sustaining Afforestation Strategies," in M. Cernea, ed. *Putting People First: Sociological Variables in Rural Development*, New York: Oxford University Press, pp. 267–93.

Chambers, Robert, N. C. Saxena and Tushaar Shah. 1989. *To the Hands of the Poor: Water and Trees*, Boulder: Westview Press.

Chertow, Marian R. and Daniel C. Esty, eds. 1997. *Thinking Ecologically: The Next Generation of Environmental Policy*, New Haven: Yale University Press.

Church, Thomas and Robert Nakamura. 1993. *Cleaning Up the Mess: Implementation Strategies in Superfund*, Washington, DC: Brookings Institution.

Clarke, J. N. and D. McCool. 1985. *Staking Out the Terrain: Power Differentials among Natural Resource Management Agencies*, Albany: State University of New York Press.

Clay, Jason. 1988. *Indigenous Peoples and Tropical Forests: Models of Land Use and Management from Latin America*, Cambridge, Mass.: Institute for Cultural Survival.

Clifford, Sue. 1998. "Halcyon Days," in D. Warburton, ed. *Community and Sustainable Development: Participation in the Future*, London: Earthscan.

Cochrane, Glynn. 1979. *The Cultural Appraisal of Development Projects*, New York: Praeger.

Cohen, Jack and Ian Stewart. 1994. *The Collapse of Chaos*, London: Viking/Penguin.

Cortner, Hanna J. And Margaret A. Moote. 1999. *The Politics of Ecosystem Management*, Covelo, Calif.: Island Press.

Costanza, Robert, Bryan Norton and Benjamin Haskell, eds. 1992. *Ecosystem Health: New Goals for Environmental Management*, Covelo, Calif.: Island Press.

Cronon, William. 1991. Nature's Metropolis: Chicago and the Great West, New York: W.W. Norton.

Cronon, William, George Miles and Jay Gitlin. 1992. *Under an Open Sky: Rethinking America's Western Past*, New York: W.W. Norton.

Crowfoot, James E. and Julia M. Wondolleck. 1990. *Environmental Disputes: Community Involvement in Conflict Resolution*, Covelo, Calif.: Island Press.

Daly, Herman E. 1996. *Beyond Growth: The Economics of Sustainable Development*, Boston: Beacon Press.

Daily, Gretchen, ed. 1997. *Nature's Services: Societal Dependence on Natural Ecosystems*, Covelo, Calif.: Island Press.

Daniel, J. C. and J. S. Serrao, eds. 1990. *Conservation in Developing Countries: Problems and Prospects*, Bombay: Oxford University Press.

Daniels, Tom. 1999. *When City and Country Collide: Managing Growth in the Metropolitan Fringe*, Covelo, Calif.: Island Press.

Dauvergne, Peter. 1997. *Shadows in the Forest: Japan and the Politics of Timber in Southeast Asia*, Cambridge, Mass.: MIT Press.

Davis, Kingsley and Kikhail S. Bernstam, eds. 1991. *Resources, Environment and Population: Present Knowledge, Future Options*, New York: Oxford University Press.

Denslow, Julie Sloan and Christine Padoch, eds. 1988. *People of the Tropical Rain Forest*, Berkeley: University of California Press/Smithsonian Institution.

Dobkowski, Michael and Isidor Walliman, eds. 1997. *The Coming Age of Scarcity: Preventing Mass Death and Genocide in the Twenty-First Century*, Syracuse: Syracuse University Press.

Doppelt, Bob, Mary Scurlock, Chris Frissel and James Karr. 1993. *Entering the Watershed: A New Approach to Save America's River Ecosystems*, Washington, DC and Covelo, Calif.: Island Press.

Douglas, Mary. 1990. "Converging on Autonomy: Anthropology and Institutional Economics," in Oliver Williamson, ed. *Organization Theory: From Chester Barnard to the Present and Beyond*, New York and Oxford: Oxford University Press.

Dowie, Mark. 1995. *Losing Ground: American Environmentalism at the Close of the Twentieth Century*, Cambridge, Mass.: MIT Press.

Drengson, Alan and Duncan Taylor, eds. 1997. *Ecoforestry: The Art and Science of Sustainable Forest Use*, Gabriola Island, BC: New Society Publishers.

Drury Jr., William Holland, 1998. *Chance and Change: Ecology for Conservationists*, Berkeley: University of California Press.

Dudley, Nigel, Jean-Paul Jeanrenaud and Francis Sullivan. 1995. *Bad Harvest? The Timber Trade and the Degradation of the World's Forests*, London: Earthscan.

Dye, Thomas R. 1972. *Understanding Public Policy*, Englewood Cliffs: Prentice-Hall.

Edington, John and M. Ann Edington. 1986. *Ecology, Recreation and Tourism*, Cambridge, Mass.: Cambridge, Mass. University Press.

Edwards, Michael and David Hulme, eds. 1993. *Making a Difference: NGOs and Development in a Changing World*, London: Earthscan.

Ellis, Richard. 1993. *American Political Cultures*, New York: Oxford University Press.

Endicott, Eve, ed. 1993. *Land Conservation through PUBLIC/Private Partnerships*, Covelo, Calif.: Island Press.

Esman, Milton J. 1991. *Management Dimensions of Development: Perspectives and Strategies*, West Hartford: Kumarian Press.

Esman, Milton J. 1994. *Ethnic Politics*, Ithaca: Cornell University Press.

Falloux, Francois and Lee Talbot. 1993. *Crisis and Opportunity: Environment and Development in Africa*, London: Earthscan.

Fischer, Kurt and Johan Schot, eds. 1993. *Environmental Strategies for Industry: International Perspectives on Research Needs and Policy Implications*, Covelo, Calif.: Island Press.

Forrester, Jay W. 1973. "The Counterintuitive Nature of Social Systems," in Franklin Tugwell, ed. *Search for Alternatives: Public Policy and the Study of the Future*, Cambridge, Mass.: Winthrop, pp. 198–223.

Fox, Jonathan A. and L. David Brown, eds. 1998. *The Struggle for Accountability: The World Bank, NGOs, and the Grassroots Movements*, Cambridge, Mass.: MIT Press.

Freedman, Jonathan. 1975. *Crowding and Behavior: The Psychology of High-Density Living*, New York: Viking Press.

Freese, Curtis H. 1998. *Wild Species as Commodities: Managing Markets and Ecosystems for Sustainability*, Covelo, Calif.: Island Press.

Freyfogle, Eric T. 1998. *Bounded People, Boundless Lands: Envisioning a New Land Ethic*, Covelo, Calif.: Island Press.

Freidmann, John. 1992. *Empowerment: The Politics of Alternative Development*, Cambridge: Blackwell.

Freidmann, John and Haripriya Rangan, eds. 1993. *In Defense of Livelihood: Comparative Studies on Environmental Action*, West Hartford: Kumarian Press.

Fumento, Michael. 1993. *Science under Siege: Balancing Technology and the Environment*, New York: William Morrow and Company.

Gale, Robert and Stephen Barg with Alexander Gillies. 1995. *Green Budget Reform: An International Casebook on Leading Practices*, London: Earthscan.

Gallagher, Mark. 1991. *Rent-seeking and Economic Growth in Africa*, Boulder: Westview Press.

Gamba, Julio and others. 1986. *Industrial Energy Rationalization in Developing Countries*, Baltimore: Johns Hopkins University Press.

Ghate, Prabhu. 1992. *Informal Finance: Some Findings from Asia*, published for the Asian Development Bank, Hong Kong: Oxford University Press.

Ghimire, Krishna. 1992. *Forest or Farm?: The Politics of Poverty and Land Hunger in Nepal*, Delhi: Oxford University Press.

Goldschmidt, Tijs. 1996. *Darwin's Dreampond: Drama in Lake Victoria*, Cambridge, Mass.: MIT Press.

Goodland, Robert et al, eds. 1992. *Population, Technology and Lifestyle: The Transition to Sustainability*, Covelo, Calif.: Island Press.

Gradwohl, Judith and Russell Greenberg. 1988. *Saving the Tropical Forests*, London: Earthscan.

Grainger, Alan. 1988. "Tropical Rainforests—Global Resource or National Responsibility?" in V. Martin, ed., *For the Conservation of Earth*, Golden: Fulcrum, pp. 94–99.

Grant, Sandy and Brian Egner. 1989. "The Private Press and Democracy," in John Holm and Patrick Molutsi, eds. *Democracy in Botswana*, Gaborone: Macmillan Botswana Publishing Co., pp. 247–63.

Gray, Gary. 1993. *Wildlife and People: The Human Dimensions of Wildlife Ecology*, Urbana: University of Illinois Press.

Gregersen, Hans, et al. 1989. *People and Trees: The Role of Social Forestry in Sustainable Development*, Washington, DC: World Bank.

Grindle, Merilee. 1977. *Bureaucrats, Politicians and Peasants in Mexico: A Case Study in Public Policy*, Berkeley: University of California Press.

Grindle, Merilee, ed. 1980. *Politics and Policy Implementation in the Third World*,

Princeton: Princeton University Press.

Grindle, Merilee and John Thomas. 1991. *Public Choices and Policy Change: The Political Economy of Reform in Developing Countries*, Baltimore: Johns Hopkins University Press.

Grubb, Michael, Matthias Koch, Abby Munson, Francis Sullivan and Koy Thomson. 1993. *The Earth Summit Agreements: A Guide and Assessment*, London: Earthscan.

Grumbine, R. Edward, ed. 1994. *Environmental Policy and Biodiversity*, Covelo, Calif.: Island Press.

Grzimek, Bernhard. 1970. *Among Animals of Africa*, New York: Stein and Day.

Haas, Peter. 1990. *Saving the Mediterranean: The Politics of International Environmental Cooperation*, New York: Columbia University Press.

Haas, Peter, Robert O. Keohane and Marc Levy. 1993. *Institutions for the Earth: Sources of Effective International Environmental Protection*, Cambridge, Mass.: MIT Press.

Hajer, Maarten A. 1995. *The Politics of Environmental Discourse: Ecological Modernization and the Policy Process*, Oxford: Clarendon Press.

Hall, Edward T. 1976. *Beyond Culture*, New York: Anchor/Doubleday.

Hamilton, John Maxwell. 1990. *Entangling Alliances: How the Third World Shapes Our Lives*, Cabin John, MD: Seven Locks Press.

Hanke, Steve. 1987. *Privatization and Development*, San Francisco: Institute for Contemporary Studies.

Hanna, Susan, Carl Folke and Karl-Goran Maler, eds. 1996. *Rights to Nature: Ecological, Economic, Cultural, and Political Principles of Institutions for the Environment*, Covelo, Calif.: Island Press.

Hannan, Michael T. and John Freeman. 1989. *Organizational Ecology*, Cambridge, Mass.: Harvard University Press.

Hardin, Garret. 1993. *Living within Limits: Ecology, Economics and Population Taboos*, New York: Oxford University Press.

Harris, Marvin. 1974. *Cows, Pigs, Wars, and Witches: The Riddles of Culture*, New York: Random House.

Harrison, Paul. 1987. *The Greening of Africa: Breaking through in the Battle for Land and Food*, New York: Penguin.

Harrison, Paul. 1992. *The Third Revolution: Environment, Population and a Sustainable World*, London: I. B. Tauris & Co.

Hawken, Paul. 1993. *The Ecology of Commerce*, New York: Harper Business.

Hayes, Harold T. P. 1977. *The Last Place on Earth*, New York: Stein and Day.

Head, Suzanne and Robert Heinzman, eds. 1990. *Lessons of the Rainforest*, San Francisco: Sierra Club Books.

Hecht, Susanna and Alexander Cockburn. 1990. *The Fate of the Forest: Developers, Destroyers and Defenders of the Amazon*, New York: Harper.

Heginbotham, Stanley. 1975. *Cultures in Conflict: The Four Faces of Indian Bureaucracy*, New York: Columbia University Press.

Heilbroner, Robert L.1975. *An Inquiry into the Human Prospect*, New York: W.W. Norton.

Hirschhorn, Joel and Kirsten Oldenburg. 1991. *Prosperity without Pollution: The Prevention Strategy for Industry and Consumers*, New York: Van Nostrand Reinhold.

Hirschman, Albert O. 1967. *Development Projects Observed*, Washington, DC: Brookings Institution.

Hirschman, Albert O. 1970. *Exit, Voice and Loyalty: Responses to Decline in Firms, Organizations and States*, Cambridge, Mass.: Harvard University Press.

Hitchcock, Robert K. 1997. 'African Wildlife: Conservation and Conflict,' in B. R. Johnston, ed. *Life and Death Matters: Human Rights and the Environment at the End of the Millennium*, London: Altamira Press, pp. 80–95.

Holman, Dennis. 1967. *The Elephant People*, London: John Murray.

Holmberg, Johan, ed. 1992. *Making Development Sustainable: Redefining Institutions, Policy and Economics*, Covelo, Calif.: Island Press.

Honadle, Beth W. 1983. *Public Administration in Rural Areas and Small Jurisdictions: A Guide to the Literature*, New York: Garland Publishing Company.

Honadle, George. 1993. "Institutional Constraints on Sustainable Resource Use: Lessons from the Tropics Showing That Resource Overexploitation is Not Just an Attitude Problem and Conservation Education is Not Enough," in Gregory Aplet, Nels Johnson, Jeffrey T. Olson and V. Alaric Sample, eds. *Defining Sustainable Forestry*, Covelo, Calif.: Island Press, pp. 90–119.

Honadle, George and Lauren Cooper. 1990. "Closing the Loops: Workshop Approaches to Evaluating Development Projects," in K. Finsterbusch, J. Ingersoll and L. Llewellyn, eds. *Methods for Social Analysis in Developing Countries*, Boulder: Westview Press, pp. 185–202.

Honadle, George and Rudi Klauss, eds. 1979. *International Development Ad-*

ministration: *Implementation Analysis for Development Projects*, New York: Praeger.

Honadle, George and Jerry Vansant. 1985. *Implementation for Sustainability: Lessons from Integrated Rural Development*, West Hartford: Kumarian Press.

Horton, Tom. 1987. *Bay Country*, Baltimore: Johns Hopkins University Press.

Hughes, Francine. 1987. "Conflicting Uses for Forest Resources in the Lower Tana River Basin of Kenya," in D. Anderson and R. Grove, eds. *Conservation in Africa: People, Policies and Practices*, Cambridge, Mass.: Cambridge, Mass. University Press, pp. 211–28.

Hurst, Philip. 1990. *Rainforest Politics: Ecological Destruction in Southeast Asia*, London: Zed Books.

Hyden, Goran. 1983. *No Shortcuts to Progress: African Development Management in Perspective*, Berkeley: University of California Press.

Jacobs, Louis. 1993. *Quest for the African Dinosaurs*, New York: Villard Books.

Jepma, C. J. 1995. *Tropical Deforestation: A Socio-economic Approach*, London: Earthscan.

Jesudason, James. 1990. *Ethnicity and the Economy: The State, Chinese Business, and Multinationals in Malaysia*, Singapore: Oxford University Press.

Joaquin, Nick. 1972. *The Woman Who Had Two Navels*, Manila: Solidaridad Publishing House.

Johannes, Robert, ed. 1989. *Traditional Ecological Knowledge: A Collection of Essays*, Gland, Switzerland: International Union for the Conservation of Nature.

Johnston, Barbara Rose, ed. 1994. *Who Pays the Price?: The Sociocultural Context of Environmental Crisis*, Covelo, Calif.: Island Press.

Jordan, Carl. 1986. "Local Effects of Tropical Deforestation," in M. Soule, ed. *Conservation Biology*, Sunderland, Mass: Sinaur, pp. 410–26.

Kamath, Shyam J. 1992. *The Political Economy of Suppressed Markets: Controls, Rent-seeking and Interest-group Behaviour in the Indian Sugar and Cement Industries*, Delhi: Oxford University Press.

Kirkby, John, Phil O'Keefe and Lloyd Timberlake. 1995. *The Earthscan Reader in Sustainable Development*, London: Earthscan.

Klee, Gary, ed. 1980. *World Systems of Traditional Resource Management*, New York: Halsted Press.

Knight, Richard L. and Peter B. Landres, eds. 1998. *Stewardship across Bound-*

aries, Covelo, Calif.: Island Press.

Korten, David C, ed. 1986. *Community Management: Asian Experience and Perspectives*, West Hartford: Kumarian Press.

Korten, David C. 1988. 'From Bureaucratic to Strategic Organization,' in F. Korten and R. Siy, eds. Transforming a Bureaucracy: *The Experience of the Philippine National Irrigation Administration*, West Hartford: Kumarian Press, pp.117–44.

Korten, David C. 1991. *Getting to the Twenty-First Century: Voluntary Action and the Global Agenda*, West Hartford: Kumarian Press.

Korten, David C. 1999. *The Post-corporate World: Life after Capitalism*, West Hartford: Kumarian Press.

Korten, Frances F. 1986. "The Policy Framework for Community Management," in D. C. Korten, ed. *Community Management: Asian Experience and Perspectives*, West Hartford: Kumarian Press.

Korten, Frances F. 1988. "The Working Group as a Catalyst for Organizational Change," in F. Korten and R. Siy, eds. *Transforming a Bureaucracy: The Experience of the Philippine National Irrigation Administration*, West Hartford: Kumarian Press, pp. 61–89.

Korten, Frances F. and Robert Y. Siy, Jr., eds. 1988. *Transforming a Bureaucracy: The Experience of the Philippine National Irrigation Administration*, West Hartford: Kumarian Press.

Korten, Frances F. and Robert Y. Siy, Jr. 1988. "Summary and Conclusions," in F. Korten and R. Siy, eds. *Transforming a Bureaucracy: The Experience of the Philippine National Irrigation Administration*, West Hartford: Kumarian Press, pp. 145–57.

Kotkin, Joel. 1993. *Tribes: How Race, Religion and Identity Determine Success in the New Global Economy*, New York: Random House.

Krueger, Anne O. 1992. *A Synthesis of the Political Economy in Developing Countries; volume 5 of The Political Economy of Agricultural Pricing Policy*, Baltimore: Johns Hopkins University Press.

LaMay, Craig and Everette Dennis. eds. 1991. *Media and the Environment*, Covelo, Calif.: Island Press.

Landy, Marc, Marc Roberts and Stephen Thomas. 1990. *The Environmental Protection Agency: Asking the Wrong Questions*, New York: Oxford University Press.

Ledec, George. 1985. "The Political Economy of Tropical Deforestation," in H.

J. Leonard, ed., *Divesting Nature's Capital*, New York: Holmes and Meier, pp. 179–226.

Ledec, George and Robert Goodland. 1988. *Wildlands: Their Protection and Management in Economic Development*, Washington, DC: World Bank.

Leonard, David. 1991. *African Successes: Four Public Managers of Kenyan Rural Development*, Berkeley: University of California Press.

Leonard, H. Jeffrey, ed. 1985. *Divesting Nature's Capital: The Political Economy of Environmental Abuse in the Third World*, New York: Holmes and Meier.

Leonard, H. Jeffrey and others. 1989. *Environment and the Poor: Development Strategies for a Common Agenda*, Washington, DC: Overseas Development Council.

Lewis, Martin W. 1992. *Green Delusions: An Environmentalist Critique of Radical Environmentalism*, Durham: Duke University Press.

Lindenberg, Marc and Benjamin Crosby. 1981. *Managing Development: The Political Dimension*, West Hartford: Kumarian Press.

Liphuko, Seeiso D. 1989. "Civil Service Consultation: An Examination of Three Cases," in J. Holm and P. Molutsi, eds. *Democracy in Botswana*, Gaborone: Macmillan Botswana Publishing Co., pp. 231–37.

Litfin, Karen T., ed. 1998. *The Greening of Sovereignty in World Politics*, Cambridge, Mass.: MIT Press.

Little, Peter and David Brokenshaw. 1987. "Local Institutions, Tenure and Resource Management in East Africa," in D. Anderson and R. Grove, eds. *Conservation in Africa: People, Policies and Practice*, Cambridge, Mass.: Cambridge, Mass. University Press, pp. 193–210.

Lovelace, George W. and A. Terry Rambo. 1991. "Behavioral and Social Dimensions," in K. William Easter, et al., eds. *Watershed Resource Management: Studies from Asia and the Pacific*, Singapore: Institute of Southeast Asian Studies.

Lovell, Katherine. 1992. *Breaking the Cycle of Poverty: The BRAC Strategy*, West Hartford: Kumarian Press.

Maass, Arthur and Raymond Anderson. 1978 . . . .*And the Desert Shall Rejoice: Conflict, Growth and Justice in Arid Environments*, Cambridge, Mass.: MIT Press.

MacDonnell, Lawrence and Sarah Bates, eds. 1993. *Natural Resources Policy and Law: Trends and Directions*, Covelo, Calif.: Island Press.

Mander, Jerry and Edward Goldsmith, eds. 1996. *The Case **Against** the Global Economy and **For** a Turn Toward the Local*, San Francisco: Sierra Club Books.

Mantell, Michael, Stephan Harper and Luther Propst. 1990. *Creating Successful Communities: A Guidebook to Growth Management Strategies*, Covelo, Calif.: Island Press.

Margolis, Richard and Nick Salafsky. 1998. *Measures of Success: Designing, Managing and Monitoring Conservation and Development Projects*, Covelo, Calif.: Island Press.

Mather, Alexander, ed. 1993. *Afforestation: Policy, Planning and Progress*, London: Bellhaven Press.

Mazmanian, Daniel and Paul Sabatier. 1989. *Implementation and Public Policy*, Lanham, Md: University Press of America.

McKibben, Bill. 1989. *The End of Nature*, New York: Random House.

McPhee, John. 1998. *Annals of the Former World*, New York: Farrar, Straus and Giroux.

Meltz, Robert, Dwight Merriam and Richard Frank. 1999. *The Takings Issue: Constitutional Limits on Land Use Control and Environmental Regulation*, Covelo, Calif.: Island Press.

Mendes, Chico. 1989. *Fight for the Forest*, London: Latin American Bureau.

Migdal, Joel. 1988. *Strong Societies and Weak States: State-society Relations and State Capabilities in the Third World*, Princeton: Princeton University Press.

Mikesell, Raymond and Lawrence Williams. 1992. *International Banks and the Environment*, San Francisco: Sierra Club Books.

Miller, Morris. 1991. *Debt and the Environment: Converging Crises*, New York: United Nations.

Monke, Eric A. and Scott R. Pearson. 1989. *The Policy Analysis Matrix for Agricultural Development*, Ithaca: Cornell University Press.

Montgomery, John D. 1974.*Technology and Civic Life: Making and Implementing Development Decisions*, Cambridge, Mass.: MIT Press.

Moran, Emilio. 1990. "Private and Public Colonization Schemes in Amazonia," in David Goodman and Anthony Hall, eds. *The Future of Amazonia: Destruction or Sustainable Development?*, London: Macmillan Press, pp. 70–89.

Moran, Emilio. 1992. "Amazonian Deforestation: Local Causes, Global Consequences," in Lars O. Hansson and Britta Jungen, eds. *Human Responsibility and Global Change, Proceedings from the International Conference in Goteborg, 9–14 June, 1991*, Goteborg: University of Goteborg Section of Human Ecology, pp. 54–67.

Moyana, Henry V. 1984. *The Political Economy of Land in Zimbabwe*, Gweru, Zimbabwe: Mambo Press.

Murray, Colin. 1981. *Families Divided: The Impact of Migrant Labor in Lesotho*, Johannesburg: Raven Press.

Myers, Norman. 1993. *Ultimate Security: The Environmental Basis of Political Stability*, New York: W. W. Norton.

Nabhan, Gary Paul. 1997. *Cultures of Habitat: On Nature, Culture and Story*, Washington, DC: Counterpoint.

National Research Council. 1992. *Neem: A Tree for Solving Global Problems*, Washington, DC: National Academy Press.

Nations, James.1988. "Deep Ecology Meets the Developing World," in E. O. Wilson, ed. *Biodiversity*, Washington, DC: National Academy Press, pp. 79–82.

Nelson, Michael. 1973. *The Development of Tropical Lands: Policy Issues in Latin America*, Baltimore: Johns Hopkins University Press.

Noorman, Klaas Jan and Ton Schoot Uiterkamp, eds. 1998. *Green Households? Domestic Consumers, Environment and Sustainability*, London: Earthscan.

Noronha, Raymond and John Spears. 1985. "Sociological Variables in Forestry Project Design," in M. Cernea, ed., *Putting People First: Sociological Variables in Rural Development*, New York: Oxford University Press, pp. 227–66.

North, Douglas. 1990. *Institutions, Institutional Change and Economic Performance*, Cambridge, Mass.: Cambridge, Mass. University Press.

North, Douglas. 1994. "Constraints on Institutional Innovation: Transaction Costs, Incentive Compatibility and Historical Considerations," in V. Ruttan, ed. *Agriculture, Environment and Health: Sustainable Development in the Twenty-First Century*, Minneapolis: University of Minnesota Press, pp. 48–70.

Olson, Mancur. 1982. *The Rise and Decline of Nations*, New Haven: Yale University Press.

Omara-Ojungu, Peter. 1992. *Resource Management in Developing Countries*, New York: Longman Scientific and Technical/John Wiley & Sons.

O'Riordan, Timothy and James Cameron, eds. 1994. *Interpreting the Precautionary Principle*, London: Earthscan.

Ornstein, Robert and Paul Ehrlich. 1989. *New World New Mind*, New York: Simon & Schuster.

Orr, David W. 1992. *Ecological Literacy: Education and the Transition to a*

Postmodern World, Albany: State University of New York Press.

Ostrom, Elinor. 1987. "Institutional Arrangements for Resolving the Commons Dilemma: Some Contending Approaches," in B. M. McKay and J. M. Anderson eds., *The Question of the Commons: The Culture and Ecology of Communal Resources*, Tucson: University of Arizona Press, pp. 250–65.

Ostrom, Elinor. 1990. *Governing the Commons: The Evolution of Institutions for Collective Action*, Cambridge, Mass.: Cambridge University Press.

Ostrom, Elinor. 1992. *Crafting Institutions for Self-Governing Irrigation Systems*, San Francisco: Institute for Contemporary Studies.

Ostrom, Elinor, Larry Schroeder and Susan Wynne. 1993. *Institutional Incentives and Sustainable Development: Infrastructure Policies in Perspective*, Boulder: Westview Press.

Owens, Owen D. 1993. *Living Waters: How to Save Your Local Stream*, New Brunswick: Rutgers University Press.

Panayotou, Theodore. 1993. *Green Markets: The Economics of Sustainable Development*, Cambridge, Mass.: San Francisco: Institute for Contemporary Studies.

Park, Chris C. 1992. *Tropical Rainforests*, New York: Routledge.

Pearce, David W. and Jeremy J. Warford. 1993. *World without End: Economics, Environment and Sustainable Development*, New York: Oxford University Press.

Peffer, Randall. 1979. *Watermen*, Baltimore: Johns Hopkins University Press.

Peters, Thomas and Robert Waterman, Jr. 1982. *In Search of Excellence: Lessons from America's Best-run Companies*, New York: Warner Books.

Peters, William and Leon Neuenschwander. 1988. *Slash and Burn: Farming in the Third World Forest*, Moscow: University of Idaho Press.

Pirsig, Robert. 1974. *Zen and the Art of Motorcycle Maintenance*, New York: William Morrow.

Poffenberger, Mark. 1990. *Keepers of the Forest: Land Management Alternatives in Southeast Asia*, West Hartford: Kumarian Press.

Ponting, Clive. 1991. *A Green History of the World: The Environment and the Collapse of Great Civilizations*, New York: St. Martin's Press.

Popkin, Samuel L. 1979. *The Rational Peasant: The Political Economy of Rural Society in Vietnam*, Berkeley: University of California Press.

Porter, Michael E. 1990. *The Competitive Advantage of Nations*, New York: The Free Press.

Power, Thomas Michael. 1996. *Lost Landscapes and Failed Economies: The Search for a Value of Place*, Covelo, Calif.: Island Press.

Pressman, Jeffrey and Aaron Wildavsky.1973. *Implementation*, Berkeley: University of California Press.

Psihoyos, Louie, with John Knoebber. 1994. *Hunting Dinosaurs*, New York: Random House.

Putnam, Robert D. 1993. *Making Democracy Work: Civic Traditions in Modern Italy*, Princeton: Princeton University Press.

Reed, David, ed. 1992. *Structural Adjustment and the Environment*, Boulder: Westview Press.

Reid, Walter V. and others. 1993. *Biodiversity Prospecting: Using Genetic Resources for Sustainable Development*, Washington, DC: World Resources Institute.

Repetto, Robert and Malcolm Gillis, eds. 1988. *Public Policies and the Misuse of Forest Resources*, Cambridge, Mass.: Cambridge, Mass. University Press.

Rich, Bruce. 1994. *Mortgaging the Earth: The World Bank, Environmental Impoverishment, and the Crisis of Development*, Boston: Beacon Press.

Richardson, Boyce. 1976. *Strangers Devour the Land: A Chronicle of the Assault upon the Last Coherent Hunting Culture in North America, the Cree Indians of Quebec, and Their Vast Primeval Homelands*, New York: Knopf.

Riggs, Fred W. 1964. *Administration in Developing Countries: The Theory of Prismatic Society*, Boston: Houghton Mifflin Co.

Ringquist, Evan. 1993. *Environmental Protection at the State Level: Politics and Progress in Controlling Pollution*, Armonk, NY: M. E. Sharpe.

Robertson, James. 1990. *Future Wealth: A New Economics for the Twenty-First Century*, New York: The Bootstrap Press.

Rodda, Annabel. 1991. *Women and the Environment*, London: Zed Books.

Romm, Jeff. 1986. "Frameworks for Governmental Choice," in David Korten, ed., *Community Management: Asian Experience and Perspectives*, West Hartford: Kumarian Press, pp. 225–37.

Rondinelli, Dennis. 1983. *Development Projects as Policy Experiments: An Adaptive Approach to Development Administration*, London: Methuen.

Roodman, David Malin. 1998. *The Natural Wealth of Nations: Harnessing the Market for the Environment,* New York: W.W. Norton.

Runge, C. Ford. 1997. 'Environmental Protection from Farm to Market,' in M.

Chertow and D. Esty, eds. *Thinking Ecologically: The Next Generation of Environmental Policy*, New Haven: Yale University Press, pp. 200–16.

Russell, Clifford and Norman Nicholson, eds. 1981. *Public Choice and Rural Development*, Washington, DC: Resources for the Future.

Ruttan, Vernon W., ed. 1994. *Agriculture, Environment and Health: Sustainable Development in the 21st Century*, Minneapolis: University of Minnesota Press.

Sandford, Stephen. 1983. *Management of Pastoral Development in the Third World*, London: John Wiley & Sons.

Sargent, Frederick, Paul Lusk, Jose Rivera and Maria Varela. 1991. *Rural Environmental Planning for Sustainable Communities*, Covelo, Calif.: Island Press.

Savory, Allan. 1988. *Holistic Resource Management*, Covelo, Calif.: Island Press.

Schmidheiny, Stephan. 1992. *Changing Course: A Global Perspective on Development and the Environment*, Cambridge, Mass.: MIT Press.

Schmidheiny, Stephan and Federico Zorraquin. 1996. *Financing Change: The Financial Community, Eco-efficiency, and Sustainable Development*, Cambridge, Mass.: MIT Press.

Schramm, Gunter and Jeremy Warford, eds. 1989. *Environmental Management and Economic Development*, Baltimore: Johns Hopkins University Press.

Schumacher, E. F. 1973. *Small is Beautiful: Economics as if People Mattered*, New York: Harper and Row.

Scott, James C. 1976. *The Moral Economy of the Peasant: Rebellion and Subsistence in Southeast Asia*, New Haven: Yale University Press.

Sexton, Ken, Alfred Marcus, K. William Easter and Timothy Burkhardt, eds. 1999. *Better Environmental Decisions: Strategies for Governments, Businesses and Communities*, Covelo, Calif.: Island Press.

Sharma, Narenda, ed. 1992. *Managing the World's Forests: Looking for Balance between Conservation and Development*, Dubuque: Kendall/Hunt.

Shiva, Vandana, Patrick Anderson, Heffa Schucking, Andrew Gray, Larry Lohmann, and David Cooper. 1991. *Biodiversity: Social and Ecological Perspectives*, London and Penang: Zed Books and World Rainforest Movement.

Sinclair, A. R. E. and Michael P. Wells. 1989. "Population Growth and the Poverty Cycle: Colliding Ecological and Economic Processes?" in David Pimentel and Carl Hall, eds. *Food and Natural Resources*, San Diego: Academic Press, pp. 439–84.

Skinner, B. F. 1971. *Beyond Freedom and Dignity*, New York: Alfred A. Knopf.

Smith, Katie and Tetsunao Yamamori, eds. 1992. *Growing Our Future: Food Security and the Environment*, West Hartford: Kumarian Press.

Smith, Zachary A. 1992. *The Environmental Policy Paradox*, Englewood Cliffs: Prentice-Hall.

Spears, John. 1988. "Preserving Biological Diversity in the Tropical Forests of the Asian Region," in E. O. Wilson, ed., *Biodiversity*, Washington, DC: National Academy Press, pp. 393–402.

Stinchcombe, Arthur L. 1965. "Social Structure and Organizations," in James G. March, ed. *Handbook of Organizations*, Chicago: Rand McNally, pp. 142–93.

Stone, Christopher. 1993. *The Gnat Is Older Than Man: Global Environment and Human Agenda*, Princeton: Princeton University Press.

Sweet, Charles and Peter Weisel. 1979. "Process versus Blueprint Approaches to Designing Rural Development Projects," in George Honadle and Rudi Klauss, eds. *International Development Administration: Implementation Analysis for Development Projects*, New York: Praeger, pp. 127–45.

Tang, Shui Yan. 1992. *Institutions and Collective Action: Self-Governance in Irrigation*, San Francisco: Institute for Contemporary Studies.

Taylor, D. R. F. and Fiona Mackenzie, eds. 1992. *Development from Within: Survival in Rural Africa*, London: Routledge.

Terrie, Philip G. 1997. *Contested Terrain: A New History of Nature and People in the Adirondacks*, Syracuse: Syracuse University Press.

Thompson, James D. 1967. *Organizations in Action*, New York: McGraw-Hill.

Thomson, James T. 1988. "Deforestation and Desertification in Twentieth-Century Arid Sahelian Africa," in J. Richards and R. Tucker, eds. *World Deforestation in the Twentieth Century*, Durham: Duke University Press, pp. 70–90.

Thomson, Ron. 1986. *On Wildlife "Conservation,"* Cape Town: United Publishers International.

Timberlake, Lloyd. 1985. *Africa in Crisis: The Causes, the Cures of Environmental Bankruptcy*, London: Earthscan.

Timberlake, Lloyd. 1987. *Only One Earth: Living for the Future*, New York: Sterling.

Tomasko, Robert M. 1993. *Rethinking the Corporation: The Architecture of Change*, New York: AMACOM.

Turnbull, Colin. 1972. *The Mountain People*, New York: Simon & Schuster.

Uphoff, Norman T. 1992. *Learning from Gal Oya: Possibilities for Participatory*

Development and Post-Newtonian Social Science, Ithaca: Cornell University Press.

Upton, Christopher and Stephen Bass. 1996. *The Forest Certification Handbook*, Delray Beach: St Lucie Press.

Von Weizsacker, Ernst U. and Jochen Jesinghaus. 1992. *Ecological Tax Reform: A Policy Proposal for Sustainable Development*, London: Zed Books.

Von Weizsacker, Ernst, Amory B. Lovins and L. Hunter Lovins. 1997. *Factor Four: Doubling Wealth, Halving Resource Use*, London: Earthscan.

Wackernagle, Mathis and William Rees. 1996. *Our Ecological Footprint: Reducing Human Impact on the Earth*, Gabriola Island, BC: New Society Publishers.

Warburton, Diane, ed. 1998. *Community and Sustainable Development: Participation in the Future*, London: Earthscan.

Ward, Peter. 1994. *The End of Evolution: On Mass Extinctions and the Preservation of Biodiversity*, New York: Bantam Books.

Ward. Peter. 1997. *The Call of Distant Mammoths: Why the Ice Age Mammals Disappeared*, New York: Copernicus.

Warner, William W. 1976. *Beautiful Swimmers: Watermen, Crabs and the Chesapeake Bay*, New York: Atlantic Monthly Press/Little, Brown.

Weaver, R. Kent and Bert A. Rockman, eds. 1993. *Do Institutions Matter?* Washington, DC: Brookings Institution.

Weidensaul, Scott. 1999. Living on the Wind: *Across the Hemisphere with Migratory Birds*, New York: North Point Press.

Weiner, Jonathan. 1996. *The Beak of the Finch: A Story of Evolution in Our Time*, New York: Alfred A. Knopf.

West, Bernadette, Peter M. Sandman and Michael R. Greenberg. 1995. *The Reporter's Environmental Handbook*, New Brunswick: Rutgers University Press.

Western, David, R. Michael Wright and Shirley Strum, eds. 1994. *Natural Connections: Perspectives in Community-based Conservation*, Covelo, Calif.: Island Press.

Whelan, Tensie, ed. 1991. *Nature Tourism: Managing for the Environment*, Covelo, Calif.: Island Press.

White, Louise. 1987. *Creating Opportunities for Change: Approaches to Managing Development Programs*, Boulder: Lynne Rienner.

Wildavsky, Aaron. 1995. *But Is It True? A Citizen's Guide to Environmental Health and Safety Issues*, Cambridge, Mass.: Harvard University Press.

Wilkinson, Charles F. 1992. *Crossing the Next Meridian: Land, Water and the Future of the West*, Covelo, Calif.: Island Press.

Wilkinson, Todd. 1998. *Science under Siege: The Politicians' War on Nature and Truth*, Boulder: Johnson Books.

Wilson, E. O. 1975. *Sociobiology*, Cambridge, Mass.: The Belknap Press of Harvard University Press.

Wilson, E. O. 1992. *The Diversity of Life*, Cambridge, Mass.: The Belknap Press of Harvard University Press.

Wilson, Monica and Godfrey Wilson. 1945. *The Analysis of Social Change*, Cambridge: Cambridge University Press.

Wood, Geoffrey D. and Richard Palmer-Jones et al. 1990. *The Water Sellers: A Cooperative Venture by the Rural Poor*, West Hartford: Kumarian Press.

Yeager, Rodger and Norman N. Miller. 1986. *Wildlife, Wild Death: Land Use and Survival in Eastern Africa*, Albany: State University of New York Press.

Young, Oran. 1981. *Natural Resources and the State: The Political Economy of Resource Management*, Berkeley: University of California Press.

Young, Oran. 1982. *Resource Regimes: Natural Resources and Social Institutions*, Berkeley: University of California Press.

JOURNALS AND PERIODICALS

Anheier, Helmut K. 1992. "Economic Environments and Differentiation: A Comparative Study of Informal Sector Economies in Nigeria," *World Development*, 20/11: 1573–86.

Ankney, C. Davison. 1996. "An Embarrassment of Riches: Too Many Geese," *Journal of Wildlife Management*, 60/2: 217–23.

Annis, Sheldon. 1992. "Evolving Connectedness among Environmental Groups and Grassroots Organizations in Protected Areas of Central America," *World Development*, 20/4: 587–95.

Arnold, J. E. M. 1987. "Community Forestry," *AMBIO — A Journal of the Human Environment*, xvi/2–3: 122–28.

Bandara, C. M. Madduma. 1989. "Environmental Awareness among the Most Vulnerable Communities in Developing Countries," *International Social Science Journal*, XLI/3: 441–48.

Bartlett, Albert A. 1994. "Reflections on Sustainability, Population Growth, and the Environment," *Population and Environment: A Journal of Interdiscipli-*

nary Studies, 16/1: 5–35.

Bloom, David. 1995. "International Public Opinion on the Environment," *Science*, 269/21 July: 354–58.

Brandon, Katrina Eadie and Michael Wells. 1992. "Planning for People and Parks: Design Dilemmas," *World Development*, 20/4: 557–70.

Broad, Robin. 1994. "The Poor and the Environment: Friends or Foes?" *World Development*, 22/6: 811–22.

Brooke, Elizabeth Heilman.1993. "Brazil's Nature Savior," *Nature Conservancy*, 43/3: 10–15.

Buckman, R. E. 1987. "Strengthening Forestry Institutions in the Developing World," *AMBIO—A Journal of the Human Environment*, xvi/2–3: 120–21.

Burton, Ian and Peter Timmerman. 1989. "Human Dimensions of Global Change—A Review of Responsibilities and Opportunities," *International Social Science Journal*, XLI/3: 297–314.

Cernea, Michael M. 1993. "Culture and Organization: The Social Sustainability of Induced Development," *Sustainable Development*, 1/2: 18–29.

Chambers, Robert and Melissa Leach. 1989. "Trees as Savings and Security for the Rural Poor," *World Development*, 17/3: 329–42.

Christie, Patrick, Alan T. White, and Delma Buhat. 1994. "Community-based Coral Reef Management on San Salvador Island, The Philippines," *Society and Natural Resources*, 7/2: 103–18.

Clark, Tim. 1993. "Creating and Using Knowledge for Species and Ecosystem Conservation: Science, Organizations and Policy," *Perspectives in Biology and Medicine*, 36/3: 497–525.

Cohen, Joel. 1995. "Population Growth and Earth's Human Carrying Capacity," *Science*, 269/21 July: 341–46.

Cohn, Roger. 1994. "Zambia: The People's War on Poaching," *Audubon*, 96/2: 70–84.

Costanza, Robert and Herman Daly. 1992. "Natural Capital and Sustainable Development," *Conservation Biology*, 6/1: 37–46.

Crocker, Thomas D. and Jason F. Shogren. 1994. "Transferable Risks and the Technology of Environmental Conflict," *Society and Natural Resources*, 7/2: 181–88.

Daily, Gretchen. 1995. "Restoring Value to the World's Degraded Lands," *Science*, 269/ 21 July: 350–54.

Decker, Daniel J. and Ken G. Purdey. 1988. "Toward a Concept of Wildlife Acceptance Capacity in Wildlife Management," *Wildlife Society Bulletin*, 16/1: 53–57.

Dewees, Peter. 1989. "The Woodfuel Crisis Reconsidered: Observations on the Dynamics of Abundance and Scarcity," *World Development*, 17/8: 1159–72.

Dia, Mamadou. 1991. "Development and Cultural Values in Sub-saharan Africa," *Finance and Development*, December: 10–13.

Dinerstein, Eric and Eric D. Wikramanayake. 1993. "Beyond 'Hotspots': How to Prioritize Investments to Conserve Biodiversity in the Indo-Pacific Region," *Conservation Biology*, 7/1: 53–65.

Dobson, Andrew and Joyce Poole. 1992. "Ivory: Why the Ban Must Stay!," *Conservation Biology*, 6/1: 149, 151.

Emery, F. E. and E. L. Trist. 1965. "The Causal Texture of Organizational Environments," *Human Relations*, 18: 21–32.

Eskeland, Gunnar and Emmanuel Jimenez. 1992. "Policy Instruments for Pollution Control in Developing Countries," *The World Bank Research Observer*, 7/2: 145–70.

Ferris, James M. and Shui-Yan Tang. 1993. "The New Institutionalism and Public Administration: An Overview," *Journal of Public Administration Research and Theory*, 3/1: 4–9.

Fortmann, Louise. 1994. "Commentary: Testimony at the White House Forestry Conference," *Society and Natural Resources*, 7/2: 189–90.

Gemery, Laura. 1994. "Hooked on Green: How Cities around the World Battle Cars," *International Wildlife*, 24/1: 42–43.

Gow, David. 1992. "Poverty and Natural Resources: Principles for Environmental Management and Sustainable Development," *Environmental Impact Assessment Review*, 12: 49–65.

Granovetter, Mark. 1985. "Economic Action and Social Structure: The Problem of Embeddedness," *American Journal of Sociology*, 91/3: 481–510.

Greene, Stephan. 1991. "Second Thoughts about Debt Swaps," *The Chronicle of Philanthropy*, iv/2: 1, 12–14.

Goulden, Richard. 1989. "Thoughts on Change for Resource Managers," *Transactions of the Fifty-fourth North American Wildlife and Natural Resources Conference*, Washington, DC: Wildlife Management Institute, pp. 611–15.

Hardin, Garrett. 1968. "The Tragedy of the Commons," *Science*, 162: 1234–48.

Heath, Robin. 1986. "The National Survey of Outdoor Recreation in Zimbabwe," *Zambezia*, 13/1: 25–42.

Heberlein, Thomas. 1988. "Improving Interdisciplinary Research: Integrating the Social and Natural Sciences," *Society and Natural Resources*, 1/1: 5–16.

Heinen, Joel T. 1994. "Emerging, Diverging and Converging Paradigms on Sustainable Development," *International Journal of Sustainable Development and World Ecology*, 1/1: 22–33.

Heinen, Joel T. 1995. "Thoughts and Theory on Incentive-based Endangered Species Conservation in the United States," *Wildlife Society Bulletin*, 23/3: 338–45.

Heinen, Joel T. 1996. "Human Behavior, Incentives, and Protected Area Management," *Conservation Biology*, 10/2: 681–84.

Heinen, Joel T. and Roberta S. Low. 1992. "Human Behavioral Ecology and Environmental Conservation," *Environmental Conservation*, 19/2: 105–16.

Hoare, Richard and Johan du Toit. 1999. "Coexistence between People and Elephants in African Savannas," *Conservation Biology*, 13/3: 633–39.

Hoerner, J. Andrew. 1998. "Surveying Environmental Tax Provisions in the States: Rationales and Research Priorities," *National Tax Association Proceedings—1997*, Washington, DC: National Tax Association.

Holling, C. S. and Gary Meffe. 1996. "Command and Control and the Pathology of Natural Resource Management," *Conservation Biology*, 10/2: 328–37.

Honadle, Beth W. 1993. "Rural Development Policy: Breaking the Cargo Cult Mentality," *Economic Development Quarterly*, 7/3: 227–36.

Honadle, George. 1982a. "Rapid Reconnaissance for Development Administration: Mapping and Moulding Organizational Landscapes," *World Development*, 10/8: 633–49.

Honadle, George. 1982b. "Supervising Agricultural Extension: Practices and Procedures for Improving Field Performance," *Agricultural Administration*, 9: 29–45.

Honadle, George and Lauren Cooper. 1989. "Beyond Coordination and Control: An Interorganizational Approach to Structural Adjustment, Service Delivery and Natural Resource Management," *World Development*, 17/10: 1531–41.

Honadle, George. 1989. "Interorganizational Cooperation for Natural Resource Management: New Approaches to a Key Problem Area," *Transactions of the 54th North American Wildlife and Natural Resources Conference*, Washington, DC: Wildlife Management Institute.

Horta, Korinna. 1991. "The Last Big Rush for the Green Gold: The Plundering of Cameroon's Rainforests," *The Ecologist*, 21/3: 1–6.

Hough, John L. 1994. "Institutional Constraints to the Integration of Conservation and Development: A Case Study from Madagascar," *Society and Natural Resources*, 7/2: 119–24.

Hunter, Malcolm, Robert Hitchcock and Barbara Wyckoff-Baird. 1990. "Women and Wildlife in Southern Africa," *Conservation Biology*, 4/4: 448–51.

James, Jeffrey and Efraim Gutkind. 1985. "Attitude Change Revisited: Cognitive Dissonance Theory and Development Policy," *World Development*, 13/10–11: 1139–49.

Kandell, Jonathan. 1993. "Undamming the World Bank," *Audubon*, 95/3: 106–12.

Kaus, Andrea. 1993. "Environmental Perceptions and Social Relations in the Mapimi Biosphere Reserve," *Conservation Biology*, 7/2: 398–406.

Kiggundu, Moses, Jan Jorgensen and Taieb Hafsi. 1983. "Administrative Theory and Practice in Developing Countries: A Synthesis," *Administrative Science Quarterly*, 28/3: 66–84.

Korten, David. 1980. "Community Organization and Rural Development: A Learning Process Approach," *Public Administration Review*, 40/5: 480–511.

Korten, David. 1991. "Sustainable Development," *World Policy Journal*, 8/3 (winter 1991–92): 157–90.

Landa, Janet T. 1981. "A Theory of the Ethnically Homogeneous Middleman Group: An Institutional Alternative to Contract Law," *The Journal of Legal Studies*, x: 349–62.

Larson, Bruce and Daniel W. Bromley. 1991. "Natural Resource Prices, Export Policies, and Deforestation: The Case of Sudan," *World Development*, 19/10: 1289–97.

Laurence, William F. 1999. "Gaia's Lungs: Are Rainforests Inhaling Earth's Excess Carbon Dioxide?," *Natural History*, 108/2: 96.

Leonard, D. K., J. M. Cohen and T. C. Pinkney. 1983. "Budgeting and Financial Management in Kenya's Agricultural Ministries," *Agricultural Administration*, 14: 105–20.

Low, Bobbi S. and Joel T. Heinen. 1993. "Population, Resources and Environment: Implications of Human Behavioral Ecology for Conservation," *Population and Environment: A Journal of Interdisciplinary Studies*, 15/1: 7–41.

Lutz, Ernst and Michael Young. 1992. "Integration of Environmental Concerns into Agricultural Policies of Industrial and Developing Countries," *World Development*, 20/2: 241–53.

Matland, Richard. 1995. "Synthesizing the Implementation Literature: The Ambiguity-Conflict Model of Policy Implementation," *Journal of Public Administration Research and Theory*, 5/2: 145–74.

McCarthy, John and Yus Rusila Noor. 1996. "Bird Hunting in Krankeng, West Java: Linking Conservation and Development," *Journal of Environment & Development*, 5/1: 87–100.

McGranahan, Gordon. 1991. "Fuelwood, Subsistence Foraging, and the Decline of Common Property," *World Development*, 19/10: 1275–87.

Meyer, Carrie. 1992. "A Step Back as Donors Shift Institution Building from the Public to the Private Sector," *World Development*, 20/8: 1115–26.

Moore, Curtis. 1994. "Greenest City in the World," *International Wildlife*, 24/1: 38–43.

Moser, Michael. 1989. "Recent Successes in International Wetland Conservation," *Transactions of the Fifty-fourth North American Wildlife and Natural Resources Conference*, Washington, DC: Wildlife Management Institute, pp. 75–80.

Morgan, Elizabeth, Grant Power and Van Weigel. 1993. "Thinking Strategically about Development: A Typology of Action Programs for Global Change," *World Development*, 21/12: 1913–30.

Myers, Norman. 1995. "Environmental Unknowns," *Science*, 269/21 July: 358–60.

Peluso, Nancy Lee, Craig R. Humphrey and Louise P. Fortmann. 1994. "The Rock, the Beach and the Tidal Pool: People and Poverty in Natural Resources—Dependent Areas," *Society and Natural Resources*, 7/1: 23–38.

Pimm, Stuart, Gareth Russell, John Gittleman, and Thomas Brooks. 1995. "The Future of Biodiversity," *Science*, 269: 347–50.

Radelet, Steven. 1992. "Reform without Revolt: The Political Economy of Economic Reform in the Gambia," *World Development*, 20/8: 1087–99.

Redclift, Michael. 1992. "A Framework for Improving Environmental Management: Beyond the Market Mechanism," *World Development*, 20/2: 255–59.

Redford, Kent H. and Allyn Maclean Stearman, 1993. "Forest-dwelling Native Amazonians and the Conservation of Biodiversity: Interests in Common or in Collision?," *Conservation Biology*, 7/2: 248–55.

Repetto, Robert. 1987. "Creating Incentives for Sustainable Forest Development," *AMBIO—A Journal of the Human Environment*, xvi/2-3: 94–99.

Robinson, John G. 1993. "The Limits of Caring: Sustainable Living and the Loss of Biodiversity," *Conservation Biology*, 7/1: 20–28.

Robinson, John G., Kent H. Redford and Elizabeth L. Bennett. 1999. "Wildlife Harvest in Logged Tropical Forests," *Science*, 284: 595–96.

Rose, Debra. 1992. "Free Trade and Wildlife Trade," *Conservation Biology*, 6/1: 148, 150.

Salafsky, Nick, Barbara L. Dugelby and John W. Terborgh. 1993. "Can Extractive Reserves Save the Rainforest? An Ecological and Socioeconomic Comparison of Nontimber Forest Extraction Systems in Peten, Guatemala and West Kalimantan, Indonesia," *Conservation Biology*, 7/1: 39–52.

Schmidt, Gregory. 1992. "Beyond the Conventional Wisdom: USAID Projects, Interorganizational Linkages, and Institutional Reform in Peru," *The Journal of Developing Areas*, 26: 431–56.

Schwartzman, Stephan. 1991. "Deforestation and Popular Resistance in Acre: From Local Social Movement to Global Network," *The Centennial Review*, xxxv/2: 397–422.

Schwartzman, Stephan. 1992. "Land Distribution and the Social Costs of Frontier Development in Brazil: Social and Historical Context of Extractive Reserves," *Advances in Economic Botany*, 9: 51–66.

Sharma, Shalendra D. 1996. "Building Effective International Environmental Regimes: The Case of the Global Environmental Facility," *Journal of Environment & Development*, 5/1: 73–86.

Sjoberg, Lennart. 1989. "Global Change and Human Action: Psychological Perspectives," *International Social Science Journal*, XLI/3: 413–32.

Southgate, Douglas, Rodrigo Sierra and Lawrence Brown. 1991. "The Causes of Tropical Deforestation in Ecuador: A Statistical Analysis," *World Development*, 19/9: 1145–51.

Tilman, D., D. Wedin and J. Knops. 1996. "Productivity and Sustainability Influenced by Biodiversity in Grasslands Ecosystems," *Nature*, 379: 718–20.

Tobey, James A. 1993. "Toward A Global Effort to Protect the Earth's Biological Diversity," *World Development*, 21/12: 1931–45.

Uphoff, Norman. 1993. "Grassroots Organizations and NGOs in Rural Development: Opportunities with Diminishing States and Expanding Markets," *World Development*, 21/4: 607–22.

Van Arkadie, Brian. 1989. "The Role of Institutions in Development," *Proceedings of the World Bank Annual Conference on Development Economics,* (Supplement to the *World Bank Economic Review and the World Bank Research Observer)* pp. 153–76.

Wade, Robert. 1985. "The Market for Public Office: Why the Indian State Is Not Better at Development," *World Development*, 13/4: 467–98.

Weiner, David. 1993. "The Current State of Design Craft: Borrowing, Tinkering and Problem Solving," *Public Administration Review*, 53/2: 110–20.

Wille, Chris. 1993. "Riches from the Rain Forest," *Nature Conservancy*, 43/1: 10–17.

Willers, Bill. 1994. "Sustainable Development: A New World Deception," *Conservation Biology*, 8/4: 1146–148.

Zimmerer, Karl S. 1993. "Soil Erosion and Labor Shortages in the Andes with Special Reference to Bolivia, 1953-91: Implications for 'Conservation with Development'," *World Development*, 21/10: 1659–76.

DOCUMENTS AND REPORTS

African Development Bank and Economic Commission for Africa. 1988. *Economic Report on Africa 1988*, Abidjan and Addis Ababa: ADB/ECA.

Agency For International Development. 1992. *Plan For Supporting Natural Resources Management in Sub-Saharan Africa: Regional Environmental Strategy for the Africa Bureau*, Washington, DC: AID.

Agency for International Development. 1993. *Towards a Sustainable Future for Africa, Technical Paper No. 5, Bureau for Africa, Office of Analysis, Research and Technical Support*, Washington, DC: AID.

Ahmad, Yusuf, Salah El Serafy and Ernst Lutz. 1989. *Environmental Accounting for Sustainable Development*, Washington, DC: World Bank.

Article 19—International Centre on Censorship. 1991. *Information Freedom and Censorship: World Report 1991*, Chicago: American Library Association.

Berg, Elliot J., coordinator. 1993. *Rethinking Technical Cooperation: Reforms for Capacity Building in Africa*, New York: United Nations Development Programme / Development Alternatives, Inc.

"Berg Report," see *World Bank*, 1981.

Bernstein, Janis D. 1993. *Alternative Approaches to Pollution Control and Waste Management: Regulatory and Economic Instruments*, Washington, DC: World Bank.

Bhatnagar, Bhuvan and Aubrey C. Williams, eds. 1992. *Participatory Develop-*

ment and the World Bank: Potential Directions for Change, Discussion paper no. 183, Washington, DC: World Bank.

Brinkerhoff, Derick, James Gage and Jo Anne Yeager. 1992. *Implementing Natural Resource Management Policy in Africa: A Document and Literature Review*, report prepared for AID/AFR/ARTS/FARA, Washington, DC: Abt Associates.

Brinkerhoff, Derick and James Gage. 1993. *Forestry Policy Reform in Mali: An Analysis of Implementation Issues*, report to AID, Implementing Policy Reform Project, Washington, DC: Management Systems International.

Brinkerhoff, Derick and Jo Anne Yeager, 1993. *Madagascar's Environmental Action Plan: A Policy Implementation Perspective*, report to AID, Implementing Policy Reform Project, Washington, DC: Management Systems International.

Bromley, Daniel and Michael Cernea. 1989. *The Management of Common Property Natural Resources: Some Conceptual and Operational Fallacies, Discussion Paper Number 57*, Washington, DC: World Bank.

Bryant, Coralie. 1992. "Culture, Management and Institutional Assessment," paper presented to an international conference on culture and development in Africa, Washington, DC: World Bank.

Center for the Study of Responsive Law, 1993. *Forty Ways to Make Government Purchasing Green*, Washington, DC: CSRL.

Cernea, Michael. 1989. *User Groups as Producers in Participatory Afforestation Strategies, Discussion Paper no. 70*, Washington, DC: World Bank.

Cernea, Michael. 1992. *The Building Blocks of Participation: Testing Bottom-up Planning, Discussion Paper no. 166*, Washington, DC: World Bank.

Chapman, Duane. 1993. "Environment, Income and Development in Southern Africa: An Analysis of the Interaction of Environmental and Macro Economics," *Working Paper no. 7*, Columbus: Environmental and Natural Resources Policy and Training project/Midwestern Universities Consortium for International Development.

Chua, Thia-Eng and Louise Fallon Scura, eds. 1992. *Integrative Framework and Methods for Coastal Area Management*, Manila: International Center for Living Aquatic Resources Management.

Cleaver, Kevin and Gotz Schreiber. 1993. *The Population, Agriculture and Environment Nexus in Sub-Saharan Africa (revised), Agriculture and Rural Development Series No. 9*, Washington, DC: World Bank.

Club of Dublin. 1991. *Issues Facing National Environmental Action Plans in Africa, Report from a Club of Dublin workshop, Mauritius, June 17–19, 1991,* available through the Environment Division/Africa Region/Technical Department of the World Bank.

Cook, Cynthia C. and Mikael Grut. 1989. *Agroforestry in Sub-Saharan Africa: A Farmer's Perspective, World Bank Technical Paper No. 112,* Washington, DC: World Bank.

Cooper, Lauren. 1987. "Process Consulting Approaches to Cross-cultural Development Assistance: Two Asian Cases," paper delivered to the 13th annual congress of SIETAR International, Montreal, May.

Davis, Shelton, ed. 1993. *Indigenous Views of Land and the Environment, World Bank Discussion Paper No. 188,* Washington, DC: World Bank.

Davis, Shelton and Alaka Wali. 1993. "Indigenous Territories and Tropical Forest Management in Latin America," *working paper WPS 1100,* Washington, DC: World Bank.

Dia, Mamadou. 1992. "Indigenous Management Practices: Lessons for Africa's Management in the 90's," *Concept Paper for Regional Study, Africa Technical Department/Institutional Development and Management,* Washington, DC: World Bank.

Dogse, Peter and Bernd von Droste. 1990. *Debt-for-Nature Exchanges and Biosphere Reserves: Experiences and Potential, MAB Digest no. 6,* Paris: UNESCO.

Donovan, Jerome. 1989. "Debt-for-Nature Swaps in Africa: An Early Inquiry," Bethesda, Md: Development Alternatives, Inc.

Dorm-Adzobu, Clement. 1995. *New Roots: Institutionalizing Environmental Management in Africa,* Washington, DC: World Resources Institute.

Dorm-Adzobu, Clement, and others. 1991. *Community Institutions in Resource Management: Agroforestry by Mobisquads in Ghana, From the Ground Up Case Study No. 3,* Washington, DC: World Resources Institute.

Dorm-Adzobu, Clement, and others. 1991. *Religious Beliefs and Environmental Protection: The Mashegu Sacred Grove in Northern Ghana, From the Ground Up Case Study No. 4,* Washington, DC: World Resources Institute.

Duda, Mark Damian. 1987. *Floridians and Wildlife: Sociological Implications for Wildlife Conservation in Florida, Technical Report no. 2,* Nongame Wildlife Program, Florida Game and Freshwater Fish Commission.

Du Toit, R. F. et al. 1984. "Wood Usage and Tree Planting in Zimbabwe's Communal Lands: A Baseline Survey of Knowledge, Attitudes and Practices," *A report to the Forestry Commission of Zimbabwe and the World Bank*, Harare: Resource Studies.

Environment Division, Technical Department, Africa Region. 1991. "Country Capacity to Conduct Environmental Assessments in Sub-Saharan Africa," *Environmental Assessment Working Paper No. 1*, Washington, DC: World Bank.

Environment Division, Technical Department, Africa Region. 1991. "Local Participation in Environmental Assessment of Projects" *Environmental Assessment Working Paper No. 2*, Washington, DC: World Bank.

Environmental Defense Fund. 1991. "Memorandum—The Conservation of Cameroon's Tropical Forests: A Test Case for the Global Environmental Facility and the New Forest Policy," Washington, DC: EDF.

Environmental Protection Council. 1991. *Ghana Environmental Action Plan*, Accra, Ghana.

Fitzgerald, Sarah. 1989. *International Wildlife Trade: Whose Business Is It?*, Washington, DC: World Wildlife Fund.

Food and Agriculture Organization of the United Nations, World Bank, World Resources Institute, and United Nations Development Programme. n.d. *The Tropical Forestry Action Plan*, Rome: FAO.

Ford, Richard, Francis Lelo, Chandida Monyadzwe and Richard Kashweeka. 1993. *Managing Resources with PRA Partnerships: A Case Study of Lesoma, Botswana*, Worcester/Njoro: Clark University Program for International Development and Egerton University PRA Project.

Gamman, John. 1990. "A Comparative Analysis of Public Policies Affecting natural resources and the Environment: Interest Group Politics in the Eastern Caribbean," *DESFIL project*, Bethesda, MD: Development Alternatives, Inc.

Gaudreau, Martha et al. 1990. "Guinea Natural Resources Management Assessment," *NRMS project*, Washington, DC: Energy/Development International.

Government of Ireland, Environmental Institute, University College Dublin and the World Bank. 1990. *National Environmental Action Plans, Proceedings of a workshop held in Dublin, Ireland December 12–14, 1990*, available through EDIAR and AFTEN, the World Bank.

Gregersen, Hans. 1990. "Key Forestry Issues Facing Developing Countries: A Focus on Policy and Socioeconomics Research Needs and Opportunities," St. Paul: University of Minnesota Forestry for Sustainable Development Program.

Gregersen, Hans and others. 1992. "Research Strategy: Forests, Water and Watershed Management Research Team," St. Paul: Environmental and Natural

Resources Policy and Training Project/University of Minnesota.

Griffin, John and others. 1999. *Study on the Development and Management of Transboundary Conservation Areas in Southern Africa*, Lilongwe, Malawi: United States Agency for International Development.

Gulhati, Ravi. 1990. *The Making of Economic Policy in Africa*, Washington, DC: The Economic Development Institute of the World Bank.

Gulliver, P. H. 1958. *Land Tenure and Social Change among the Nyakyusa*, Kampala: East African Institute of Social Research.

Gustafson, Daniel and Veronica Clifford. 1994. *Implementation of the Gambia Environmental Action Plan, report to AID, Implementing Policy Change Project*, Washington, DC: Management Systems International.

Haeuber, Richard. 1992. *A Citizen's Guide to World Bank Environmental Assessment Procedures*, Washington, DC: The Bank Information Center.

Hitchcock, Robert K. 1997. *Decentralization, Development and Natural Resource Management in the in the Northwestern Kalahari Desert, Botswana*, report to the Biodiversity Support Project, Washington, DC: Agency for International Development.

Honadle, George. 1978. "Farmer Organization for Irrigation Water Management," Washington, DC: Development Alternatives, Inc.

Honadle, George. 1980. "Demographic Pressure and African Land Law: Implications for Policy," Washington, DC: Development Alternatives, Inc.

Honadle, George. 1986. *Development Management in Africa: Context and Strategy—A Synthesis of Six Agricultural Projects, Evaluation Special Study no. 43*, Washington, DC: Agency for International Development.

Honadle, George. 1989. *Putting the Brakes on Tropical Deforestation: Some Institutional Considerations, report to AID*, Washington, DC: Management Systems International.

Honadle, George. 1994. *Botswana's National Conservation Strategy: Organizing for Implementation, Report to AID through the Implementing Policy Change Project*, Washington, DC: Management Systems International/Abt Associates/ Development Alternatives, Inc.

Honadle, George, Scott Grosse and Paul Phumpiu. 1994. *The Problem of Linear Project Thinking in a Non-linear World: Experience from the Nexus of Population, Environmental, and Agricultural Dynamics in Africa*, St. Paul: report to the Africa Bureau of AID from the Environmental and Natural Resources Policy and Training Project/University of Minnesota.

Horta, Korinna. 1993. See Udall, Lori.

Human Rights Watch. 1991. *Human Rights Watch World Report 1992—Events of 1991*, New York: Human Rights Watch.

Human Rights Watch and Natural Resources Defense Council. 1992. *Defending the Earth: Abuses of Human Rights and the Environment*, New York and Washington, DC.: HRW/NRDC.

Institute for Rural Development. 1991. "Village Eco Development: A Strategy for Ecological Regeneration and Forest Protection," *Report submitted to the Department of Forestry*, Government of Maharashtra, India, Pune: Institute for Rural Development.

IUCN/UNEP/WWF. 1991. *Caring for the Earth: A Strategy for Sustainable Living*, Gland: International Union for the Conservation of Nature.

Jamaica, Government of. 1992. *NRCA 2000 Workshop Briefing Paper*, Kingston: Natural Resources Conservation Authority.

James, R. W. 1985. *Land Law and Policy in Papua New Guinea, Monograph no. 5*, Port Moresby: Papua New Guinea Law Reform Commission.

Josiah, Scott J. 1996. *Local Action for Global Change: Expanding the Impacts of NGO Natural Resource and Rural Development Programs, Submitted in partial fulfillment of the Doctor of Philosophy in Forest Resources, University of Minnesota*, St. Paul, Minnesota.

Kikert, Sunita, John Nellis and Mary Shirley. 1992. *Privatization: The Lessons of Experience*, Washington, DC: The World Bank.

King, Dennis, Pierre Crosson and Jason Shogrun. 1992. *Use of Economic Instruments for Environmental Protection in Developing Countries*, Arlington, Va.: Environment and Natural Resources Policy and Training Project/Winrock International Environmental Alliance.

Kiss, Agnes, ed. 1990. *Living With Wildlife: Wildlife Resource Management with Local Participation in Africa*, Washington, DC: World Bank.

Korten, David and Norman Uphoff.1981. "Bureaucratic Reorientation for Participatory Rural Development," Washington, DC: National Association of Schools of Public Affairs and Administration.

Lele, Uma and Steven Stone. 1989. *Population Pressure, the Environment and Agricultural Intensification: Variations on the Boserup Hypothesis, MADIA Discussion Paper 4*, Washington, DC: World Bank.

Lindberg, Kreg. 1991. *Policies for Maximizing Nature Tourism's Ecological and Economic Benefits*, Washington, DC: World Resources Institute.

Lusigi, Walter, compiler. 1992. *Manging Protected Areas in Africa*, Paris: UNESCO—World Heritage Fund.

MacKinnon, John and Kathy et al. 1986. *Managing Protected Areas in the Tropics*, Gland: International Union for the Conservation of Nature.

Mahar, Dennis. 1989. *Government Policies and Deforestation in Brazil's Amazon Region*, Washington, DC: World Bank.

Maharashtra, Government of—Forest Department, Directorate of Social forestry, Forest Development Corporation and Maharashtra Industrial and Technical Consultancy Organization Ltd. 1991. "Role of NGOs and Women in Joint Management of Degraded Forest Areas and Village Eco Development," prepared for a workshop sponsored by the World Bank in Pune, April 11 and 12, 1991.

Mbise, Talala. 1987. "Process Consulting Approaches to Cross-cultural Development Assistance: Two African Cases," *Paper presented to the 13th annual congress of SIETAR International*, Montreal, May.

McCormick, Scott and George Honadle. 1999. *Rapid Appraisal: Botswana Natural Resources Management Project (B/NRMP)*, report to Agency for International Development/Regional Center for Southern Africa, Burlington: Associates in Rural Development.

McGaughey, Stephen E. and Hans M. Gregersen. 1988. *Investment Policies and Financing Mechanisms for Sustainable Forestry Development*, Washington, DC: Inter-American Development Bank.

McKay, Karen LeAnn and David Gow, and others. 1990. *Enhancing the Effectiveness of Governmental and Non-Governmental Partnership in Natural Resources Management*, principal report and case studies, NRMS project, Washington, DC: Energy/Development International.

McPherson, Malcolm and Steven Radelet, eds. 1992. *Economic Recovery in the Gambia: Lessons for Sub-Saharan Africa*, Cambridge, Mass.: Harvard Institute for International Development.

Minnesota Roundtable on Sustainable Development. 1998. *Investing in Minnesota's Future: An Agenda for Sustaining Our Quality of Life, A Report to the Governor*, St. Paul: Minnesota Planning/Environmental Quality Board.

Mulk, Shams Ul. 1993. "Water Resources Management—Pakistan's Experience," paper delivered to the Conference on Environmentally Sustainable Development, Washington, DC: World Bank.

Mwenya, A. N., D. M. Lewis and G. B. Kaweche. 1990. *ADMADE Policy, Back-*

ground and Future: National Parks and Wildlife Services New Administrative Management Design for Game Management Areas, Lusaka: Republic of Zambia.

National Commission on the Environment. 1993. *Choosing a Sustainable Future*, Washington, DC: Island Press.

National Media Production Center. 1977. *Vital Documents on Water Resources*, Manila: National Media Production Center.

Odell, Malcolm J., Jr., Michael Brown, Joseph Carvalho, Marea Hatziolos, Richard Edwards, Edward Karch, Anthony Pryor and Disikalala M. Gaseitsiwe. 1993. *Midterm Evaluation of the Botswana Natural Resources Management Project*, Gainesville: Tropical Research and Development.

Office of Technology Assessment, US Congress. 1984. *Technologies to Sustain Tropical Forests*, Washington, DC: US Government Printing Office.

Office of Technology Assessment, US Congress. 1988. *Grassroots Development: The African Development Foundation*, Washington, DC: US Government Printing Office.

Organization for Economic Cooperation and Development. 1989a. *Economic Instruments for Environmental Protection*, Paris: OECD.

Organization for Economic Cooperation and Development. 1989b. *Renewable Natural Resources: Economic Incentives for Improved Management*, Paris: OECD.

Organization for Economic Cooperation and Development. 1991a. *Environmental Indicators: A Preliminary Set*, Paris: OECD.

Organization for Economic Cooperation and Development. 1991b. *Environmental Labelling in OECD Countries*, Paris: OECD.

Organization for Economic Cooperation and Development. 1991c. *Environmental Policy: How To Apply Economic Instruments*, Paris: OECD.

Organization for Economic Cooperation and Development. 1991d. *The State of the Environment*, Paris: OECD.

Organization for Economic Cooperation and Development. 1992. *Market and Government Failures in Environmental Management: Wetlands and Forests*, Paris: OECD.

Organization for Economic Cooperation and Development. 1994a. *The Distributive Effects of Economic Instruments for Environmental Policy*, Paris: OECD.

Organization for Economic Cooperation and Development. 1994b. *Environment and Taxation: The Cases of the Netherlands, Sweden and the United States*,

Paris: OECD.

Poole, Peter. 1989. "Developing a Partnership of Indigenous Peoples, Conservationists and Land Use Planners in Latin America," working paper WPS 245, Washington, DC: World Bank.

Prescott-Allen, Robert and Christina. 1982. *What's Wildlife Worth?*, London: Earthscan.

Project 88. 1988. *Harnessing Market Forces to Protect Our Environment: Initiatives for the New President*, Washington, DC: Environmental Policy Institute.

Project 88—Round II. 1991. *Incentives for Action: Designing Market-Based Environmental Strategies*, Washington, DC: Environmental Policy Institute.

Rich Bruce M. 1993. "Testimony before the Subcommittee on Foreign Operations, Committee on Appropriations, United States Senate, June 15, 1993."

Route Advisory Task Force. 1998. *Shifting the Burden: Recommended Disposition of the Proposed Chisago Electric Transmission Line Project,* Report to the Minnesota Environmental Quality Board and the Administrative Law Judge, St. Paul: Environmental Quality Board.

Sayer, Jeffrey A. 1992. "Institutional Arrangements for Forest Conservation in Africa," in Kevin Cleaver and others, eds. *Conservation of West and Central African Rainforests*, World Bank Environment Paper no. 1, Washington, DC: World Bank, 310–17.

Sebastian, Harvel et al. 1991. "Burkina Faso Natural Resources Management Action Plan," Bethesda, MD: Development Alternatives, Inc.

Seubert, Chris and Karen LeAnn McKay. 1989. "Malawi Natural Resources Management Assessment," NRMS project, Bethesda, Md: Development Alternatives, Inc.

Shaikh, Asif et al. 1988. *Opportunities for Sustained Development: Successful Natural Resources Management in the Sahel, vols. I, II, III and IV.* Washington, DC: Energy/Development International.

Shanmugaratnam, Nadarajah, Trond Veld, Anne Mossige and Mette Bovin. 1992. *Resource Management and Pastoral Institution Building in the West African Sahel, World Bank Discussion Paper no. 175*, Washington, DC: World Bank.

Shepherd, Gill. 1993. *Managing the Forest Boundary: Policies and Their Effects in Two Projects in the Tropical Moist Forests of Cameroon and Madagascar,* London: Overseas Development Institute.

Silverman, Jerry M. 1990. *Public Sector Decentralization: Economic Policy Reform and Sector Investment Programs, Division Study Paper no. 1, Public Sec-*

tor Management Division, Africa Technical Department, Washington, DC: World Bank.

Silverman, Jerry M. 1987. "Process Consulting Approaches to Cross-cultural Development Assistance," *Paper presented to the 13th annual congress of SIETAR International*, Montreal, May.

Swallow, Brent M., Daniel W. Bromley and Jeffrey A. Cochrane. 1993. Authority, Governance and Incentives in African Rangelands, Policy Brief, Columbus: Environmental and Natural Resources Policy and Training Project/Midwestern Universities Consortium for International Activities.

Thomas-Slayter, Barbara et al. 1991. *Traditional Village Institutions in Environmental Management: Erosion Control in Katheka, Kenya, From the Ground Up Case Study No. 1*, Washington, DC: World Resources Institute.

Thompson, John. 1991. *Combining Local Knowledge and Expert Assistance in Natural Resource Management: Small-scale Irrigation in Kenya, From the Ground Up Case Study No. 2*, Washington, DC: World Resources Institute.

Thorpe, Beverley. 1999. *Citizens Guide to Clean Production*, Montreal: Clean Production Action.

Tukahirwa, Eldad, ed. 1992. *Environmental and Natural Resource Management Policy and Law: Issues and Options, (Volumes l and ll)*, Kampala and Washington, DC: Institute of Environment and Natural Resources, Makerere University and World Resources Institute.

Tukahirwa, Eldad and Peter Veit. 1992. *Public Policy and Legislation in Environmental Management: Terracing in Nyarurembo, Uganda, From the Ground Up Case Study No. 5*, Washington, DC: World Resources Institute.

Udall, Lori. 1993. "Testimony before the Subcommittee on International Develeopment, Finance, Trade and Monetary Policy, Committee on Banking, Finance and Urban Affairs, United States House of Representatives, May 5, 1993" (including a submission by Korinna Horta concerning the reform of the Global Environmental facility).

United Nations Development Programme. 1992. *Human Development Report 1992*, New York: Oxford University Press.

United Nations Economic and Social Commission for Asia and the Pacific. 1986. *Environmental and Socio-Economic Aspects of Tropical Deforestation in Asia and the Pacific*, Bankok: ESCAP.

Veit, Peter, Adolfo Mascarenhas and Okyeame Ampadu-Agyei. 1995. *Lessons from the Ground Up: African Development that Works*, Washington, DC: World

Resources Institute.

"Wapenhans Report," see *World Bank*, 1992d.

Warren, Marian et al.1985. *Development Management in Africa: The Case of the Lesotho Land Conservation and Range Management Project*, Washington, DC: Agency for International Development.

Weatherly, W. Paul, J. Eugene Gibson and David Callihan. 1993. *Draft Guidelines for Environmental Endowments, report to AID*, Arlington, Va.: Environment and Natural Resources Policy and Training Project/Winrock International Environmental Alliance.

Wells, Michael and Katrina Brandon. 1992. *People and Parks: Linking Protected Area Management with Local Communities*, Washington, DC: World Bank/ World Wildlife Fund / U. S. Agency for International Development.

World Bank. 1981. *Accelerated Development in Sub-Saharan Africa: An Agenda for Action*, Washington, DC: World Bank.

World Bank. 1983. *World Development Report 1983: Managing Development*, Washington, DC: World Bank.

World Bank. 1991. *Managing Development: The Governance Dimension, A Discussion Paper*, Washington, DC: World Bank.

World Bank. 1992a. *World Development Report 1992: Development and the Environment*, Washington, DC: Oxford University Press.

World Bank. 1992b. *Annual Report 1992*, Washington, DC: World Bank.

World Bank. 1992c. *Integrated Natural Resource Management, AFTEN Working Paper No. 4*, Washington, DC: World Bank.

World Bank. 1992d. "Report of the Portfolio Management Task Force," Washington, DC: World Bank.

World Bank and Inter-American Development Bank. 1986. *Sample Bidding Documents—Procurement of Goods*, Washington, DC: World Bank and Inter-American Development Bank.

World Resources Institute and International Institute for Environment and Development. *1986, 1987, 1988–89, 1990–91, 1992–93. World Resources*, New York: Basic Books.

World Resources Institute, World Conservation Union and United Nations Environment Programme. 1992. *Global Biodiversity Strategy: Guidelines for Action to Save, Study, and Use Earth's Biotic Wealth Sustainably and Equitably*, Washington, DC: WRI/IUCN/UNEP.

Young, Mike D. and Sarah A. Ryan. 1994. "Using Environmental Indicators to Promote Environmentally, Ecologically and Socially Sustainable Resource Use: A Policy-oriented Methodology," St. Paul: report prepared for the Environmental Policy and Natural Resources Training Project/University of Minnesota.

Zazueta, Aaron. 1995. *Policy Hits the Ground: Participation and Equity in Environmental Policy-Making*, Washington, DC: World Resources Institute.

Zimbabwe Trust, Department of National Parks and Wildlife Management, and Campfire Association. 1990. *People, Wildlife and Natural Resources—the CAMPFIRE Approach to Rural Development in Zimbabwe*, Harare: The Zimbabwe Trust.

Zimmerman, R. C. 1993. *Recent Reforms in Natural Resources Management in Africa: Trends in the Roles of Public-Sector Institutions, ARTS Technical paper No. 9*, Washington, DC: Agency for International Development.

INTERVIEWS AND OTHER SOURCES

Various people were interviewed at the following organizations:

Abt Associates
African Wildlife Foundation
Agency for International Development
Bank Information Center
Center for International Law and Environment
Chemonics, Inc.
Conservation International
Development Alternatives, Inc.
Environmental Defense Fund
Environmental Law Institute
Ford Foundation
Human Rights Watch
National Wildlife Federation
Nature Conservancy International Program
United Nations Development Programme
United Nations Environmental Programme
Wildlife Conservation International
World Bank
World Resources Institute
World Wildlife Fund (USA)

The following videotapes provided perspectives on issues of sustainable development and environmental policy:

Aga Khan Foundation, USA. 1987. *First Harvest*, Washington, DC: Aga Khan Foundation.

Griesinger Films. 1991. *An Introduction to Ecological Economics*, Gates Mills, Ohio: Griesinger Films.

Griesinger Films. 1993a. *Conversation for a Sustainable Society*, Gates Mills, Ohio: Griesinger Films.

Griesinger Films. 1993b. *Investing in Natural Capital*, Gates Mills, Ohio: Griesinger Films.

National Wildlife Federation. 1988. *Our Threatened Heritage*, Washington, DC: NWF.

World Bank. 1992. *Development and the Environment: A New Partnership*, Washington, DC: World Bank.

APPENDIX A:

ENVIRONMENTAL POLICY
IMPLEMENTATION CHECKLIST (EPIC)

I: PROBLEM CONTEXT **Check if assessment conducted or action taken**

A. What resources are threatened? ____

B. What human behavior contributes to the threat? ____

C. Does the behavior cause direct damage or does it damage the resource indirectly or as part of a complex set of interactions? ____

D. Why do people engage in the harmful behavior? ____

E. Is the resource or the threat mobile? ____

F. If either is mobile, what boundaries are crossed? ____

G. Are there cyclical or threshold aspects integral to the decline in the condition of the resource? ____

H. Is the resource relatively discrete or is it embedded in a matrix of resources, species, or interactions? ____

II: BEHAVIORAL OBJECTIVES

A. Who will do what differently? ____

B. What kinds of behavioral objectives will new policy promote? ____

reducing damaging behavior? ____

restoring a resource? _____

promoting new behavior? _____

preserving a resource? _____

increasing efficiency? _____

creating markets for substitutes? _____

building capacity and new decision processes? _____

replacing measurement systems? _____

Other? _____

C. Which will be targeted and which will be systemic objectives? _____

D. Preliminary identification of difficulties involved in move to new behaviors: _____

III: SOCIAL CONTEXT AND EMBEDDEDNESS:

A. Specify which dimensions of social context will affect policy development, implementation, or impact in what way:

information? _____

salience? _____

power balance? _____

process? _____

scale/infrastructure? _____

decision system? _____

B. Specify which aspects of embeddedness are likely to be important and how:

resource dependency? _____

psychological dependency? _____

fluidity? _____

IV: POLICY OPTIONS

A. What existing policies support what undesired behaviors? _____

    Regulatory policies? _____

    Direct policy incentives? _____

    Indirect policy incentives? _____

    Policies promoting self-management? _____

    How strong are any spillovers from indirect incentives? _____

B. Which of these behaviors are embedded behaviors:

    Resource dependent behavior? _____

    Psychologically dependent behavior? _____

    Fluidity—are they weakening or strengthening or otherwise changing? _____

C. What satellite stakeholders are concerned with which behaviors? _____

D. What nuclear stakeholders need to be involved? _____

E. What are the relative strengths of different nuclear stakeholders? _____

F. What command and control approaches could be used? _____

G. What self-management approaches could be used? _____

H. What direct incentives could be used? _____

I. What indirect incentives will be important or useful? _____

J. What new indirect incentives or feedback loops will be created by policy reform? _____

V: MECHANISMS TO IMPLEMENT POLICY OPTIONS

Which mechanisms could be used to achieve the policy objectives and what roles would each play in the reform effort?

A. organizational champion _____

B. bureaucratic reorientation _____

C. environmental dispute resolution _____

D. markets E. policy pronouncement _____

F. regulation _____

G. devolution _____

H. debt-for-nature swap _____

I. political exhortation and mobilization _____

J. publicity and public awareness _____

K. subsidy _____

L. taxation _____

M. trade restriction _____

N. Other? _____

VI: PROGRAM COMPONENTS / STRATEGY DEVELOPED

A. Trait-taking components _____

B. Trait-making components _____

C. Roles for nuclear stakeholders _____

D. Indicators for behavioral change _____

E. Indicators for impact on resource _____

F. Indicators tested for contextual validity _____

G. Monitoring system for

    implementation process _____

    human behavior _____

    resource condition _____

    change in problem context _____

    change in social context _____

    change in embeddedness _____

H. Contingency plans for

    insufficient impact   ____

    unanticipated impact   ____

    anticipated negative impact   ____

    other   ____

VII: REVISIT ALL ABOVE (I-VI) AND ADJUST AS NEEDED

A. Review problem statement   ____

B. Review behavioral objectives   ____

C. Review context factors   ____

D. Review policy options   ____

E. Reconsider roles of different mechanisms   ____

F. Review, revise, and refine implementation strategy   ____

G. Review monitoring plan   ____

H. Review contingency plans   ____

VIII: IMPLEMENT THE POLICY REFORM

KEY TERMS FOR USING EPIC

PROBLEM CONTEXT

- *resource-threat relationship*: whether the threat or the resource is site-bound or mobile and the directness and visibility of the resource-threat interaction;

- *boundary congruence*: whether the problem appeared in single, dual or multiple social or legal jurisdictions or in open sea commons and the relative capacities of organizations in the different units;

- *discreteness*: whether the resource is either highly interconnected with others in a system of relationships or more independent;

- *progression*: whether the deterioration of the resource is linear or whether it is part of a cyclical process or whether it may be approaching or has crossed a threshold.

SOCIAL CONTEXT

- *informational openness of the society*: the role of the local media and the control of information constitute a major consideration when weighing alternative policy options and implementation mechanisms;

- *interorganizational power balance*: implementation of environmental policies is greatly influenced by the relative strength of implementing organizations versus competitive organizations;

- *salience*: recognition of the immediacy and magnitude of a problem by influential actors affects commitment, resources, and the outcome of alternative implementation strategies;

- *process requirements*: local ways of cementing alliances and obtaining support must be used or else the implementation process will falter at times of stress;

- *scale, space and infrastructure*: the human resource base, the size and spatial characteristics of the country, and the transportation and communication infrastructure and organizational density narrow implementation alternatives;

- *resource decision system*: local authority structures and systems for making decisions about the use of a natural resource must be understood to judge the impact of different policies.

EMBEDDEDNESS

- *embeddedness*: (1) the degree to which a country, society, region, community, group or organization depends upon a particular resource for subsistence, financial, or physical support and (2) the symbolic importance attached to the resource or practice indicates how loosely or tightly embedded it is in the workings of the place and how difficult it will be to affect changes through policy reform, and (3) changes in its status can indicate the degree of fluidity and level of opportunity available to policy makers. The three constituent parts of embeddedness are thus: (a) *resource dependency*, (b) *psychological dependency*, and (c) *fluidity*.

SATELLITE STAKEHOLDER CATEGORIES

- *intellectual*—some people are committed by belief, personal identification, and reputation, or other factor to a particular definition of a problem, its solution, or to particular solutions to problems;

- *appreciative*—some actors have an interest in an issue because they simply appreciate a resource. Many may not use it but they appreciate its being there, others see the potential to generate information and a chance for learning, others may like a view or an animal for non-consumptive purposes.

- *organizational mission*—some organizations may have a charter or history that gives them stakes in a problem area (e.g., a transportation authority, a community development society, a pollution abatement agency, an agricultural technology research center, a land use commission, a zoning board, a tax reform institute) and they will want to be involved in any discussion of changes in policies affecting the area.

- *user*—the users or consumers of a good or service that could be changed in nature, source, availability, price, or timing by a policy shift will be concerned about any policy change. They have a direct stake in any outcome that produces change or perpetuates the former situation.

- *provider*—those who provide or market any good or service impacted by a policy may gain or lose from any change.

- *power*—some actors may lose sources of power or influence over others or over their own lives.

- *status*—a policy or its means of implementation may change the roles played by some people, and this can change their social, political, or professional status relative to others, and in some cases may affect their sense of identity. Those who gain status and those who lose status mthemselves stakeholders in alternative outcomes.

- *burden shifting*—many present policies allow people and organizanizations to transfer some of the cost of their operations to others. Those who are the passers of costs and those who are the receivers of burdens are all stakeholders who stand to gain or lose from change.

- *legal*—people or organizations may gain or lose legal rights or protections as a result of changes in a law, elimination of a law, or introduction of a new law.

NUCLEAR STAKEHOLDERS

- those actors who control resources needed to implement the change; and

- those actors who can mobilize resources to block it.

FOUNDATIONS FOR TRAIT-TAKING

- *Values*—are there symbols, processes, roles, psychological factors, or cultural practices that generate loyalty or respect and that could be employed in the change process?

- *Functions*—are there education, communication, production, distribution, disposal, maintenance, governance, religious, social, recreational, employment, research, or other functions that are presently being performed by people or institutions that could perform an allied, parallel, or identical function during the reform effort?

- *Structures*—are particular organizations in place to reach subsets of the population, is it necessary to follow particular channels when conducting certain types of business, are specific pre-existing institutions the only ones that can legitimately do certain things, and are any of them crucial to the implementation strategy?

- *People*—are particular leaders able to mobilize support for new departures, do some groups have more self-interest in promoting, accepting, or engaging in the reform than others, are there non-traditional power centers that are emerging as a result of salience or the untangling of a previously embedded situation, and could any of them either spearhead or support the policy change?

INDEX

ABOUT THE AUTHOR

GEORGE HONADLE is a Visiting Research Scholar and Associate of the Center for Institutional Studies at Bowling Green State University. He served as a Peace Corps Volunteer and agricultural extension officer in Malawi from 1967 to 1970. From 1976 to 1986 he was a senior staff member of Development Alternatives, Inc. He has had ninety-eight consulting assignments in twenty-seven countries with such organizations as the World Bank, World Wildlife Fund, United Nations Development Programme, African and Asian Development Banks, Agency for International Development, and the Minnesota Environmental Quality Board. He has been an advisor to various organizations and activities such as the Minnesota Demographer's Office, Great Lakes Protection Fund, Chisago Sustainability Project, and Minnesota Water Quality Management Evaluation. He has taught at the University of Minnesota, George Washington University, American University, and Syracuse University and has lectured at thirty-one colleges, institutes, and universities in eight countries. He designed the State of Minnesota's Sustainable Development Initiative, he served on the Minnesota Roundtable on Sustainable Development, and he was a founding member of the Board of Directors of the Minnesota Center for Sustainable Development. Dr. Honadle was educated at Dickinson College, the University of Edinburgh, and Syracuse University.

 Kumarian Press is dedicated to publishing and distributing books and other media that will have a positive social and economic impact on the lives of peoples living in "Third World" conditions no matter where they live.

Kumarian Press publishes books about Global Issues and International Development, such as Peace and Conflict Resolution, Environmental Sustainability, Globalization, Nongovernmental Organizations, and Women and Gender.

To receive a complimentary catalog or to request writer's guidelines call or write:

Kumarian Press, Inc.
14 Oakwood Avenue
West Hartford, CT 06119-2127
U.S.A.

Inquiries: (860) 233-5895
Fax: (860) 233-6072
Order toll free: (800) 289-2664

e-mail: kpbooks@aol.com
Internet: www.kpbooks.com